Lecture Notes in Mathematics

Vol. 217: T. J. Jech, Lectures in Set Theory with Particular Emphasis on the Method of Forcing. V, 137 pages. 1971. DM 16,-

Vol. 218: C. P. Schnorr, Zufälligkeit und Wahrscheinlichkeit. IV, 212 Seiten. 1971. DM 20,-

Vol. 219: N. L. Alling and N. Greenleaf, Foundations of the Theory of Klein Surfaces. IX, 117 pages. 1971. DM 16,-

Vol. 220: W. A. Coppel, Disconjugacy. V, 148 pages. 1971. DM 16,-

Vol. 221: P. Gabriel und F. Ulmer, Lokal präsentierbare Kategorien. V, 200 Seiten. 1971. DM 18,-

Vol. 222: C. Meghea, Compactification des Espaces Harmoniques. III, 108 pages. 1971. DM 16,-

Vol. 223: U. Felgner, Models of ZF-Set Theory. VI, 173 pages. 1971. DM 16,-

Vol. 224: Revêtements Etales et Groupe Fondamental. (SGA 1). Dirigé par A. Grothendieck XXII, 447 pages. 1971. DM 30,-

Vol. 225: Théorie des Intersections et Théorème de Riemann-Roch. (SGA 6). Dirigé par P. Berthelot, A. Grothendieck et L. Illusie. XII, 700 pages. 1971. DM 40,-

Vol. 226: Seminar on Potential Theory, II. Edited by H. Bauer. IV, 170 pages. 1971. DM 18,-

Vol. 227: H. L. Montgomery, Topics in Multiplicative Number Theory. IX, 178 pages. 1971. DM 18,-

Vol. 228: Conference on Applications of Numerical Analysis. Edited by J. Ll. Morris. X, 358 pages. 1971. DM 26,-

Vol. 229: J. Väisälä, Lectures on n-Dimensional Quasiconformal Mappings. XIV, 144 pages. 1971. DM 16,-

Vol. 230: L. Waelbroeck, Topological Vector Spaces and Algebras. VII, 158 pages. 1971. DM 16,-

Vol. 231: H. Reiter, L^1-Algebras and Segal Algebras. XI, 113 pages. 1971. DM 16,-

Vol. 232: T. H. Ganelius, Tauberian Remainder Theorems. VI, 75 pages. 1971. DM 16,-

Vol. 233: C. P. Tsokos and W. J. Padgett. Random Integral Equations with Applications to stochastic Systems. VII, 174 pages. 1971. DM 18,-

Vol. 234: A. Andreotti and W. Stoll. Analytic and Algebraic Dependence of Meromorphic Functions. III, 390 pages. 1971. DM 26,-

Vol. 235: Global Differentiable Dynamics. Edited by O. Hájek, A. J. Lohwater, and R. McCann. X, 140 pages. 1971. DM 16,-

Vol. 236: M. Barr, P. A. Grillet, and D. H. van Osdol. Exact Categories and Categories of Sheaves. VII, 239 pages. 1971. DM 20,-

Vol. 237: B. Stenström, Rings and Modules of Quotients. VII, 136 pages. 1971. DM 16,-

Vol. 238: Der kanonische Modul eines Cohen-Macaulay-Rings. Herausgegeben von Jürgen Herzog und Ernst Kunz. VI, 103 Seiten. 1971. DM 16,-

Vol. 239: L. Illusie, Complexe Cotangent et Déformations I. XV, 355 pages. 1971. DM 26,-

Vol. 240: A. Kerber, Representations of Permutation Groups I. VII, 192 pages. 1971. DM 18,-

Vol. 241: S. Kaneyuki, Homogeneous Bounded Domains and Siegel Domains. V, 89 pages. 1971. DM 16,-

Vol. 242: R. R. Coifman et G. Weiss, Analyse Harmonique Non-Commutative sur Certains Espaces. V, 160 pages. 1971. DM 16,-

Vol. 243: Japan-United States Seminar on Ordinary Differential and Functional Equations. Edited by M. Urabe. VIII, 332 pages. 1971. DM 26,-

Vol. 244: Séminaire Bourbaki - vol. 1970/71. Exposés 382-399. IV, 356 pages. 1971. DM 26,-

Vol. 245: D. E. Cohen, Groups of Cohomological Dimension One. V, 99 pages. 1972. DM 16,-

Vol. 246: Lectures on Rings and Modules. Tulane University Ring and Operator Theory Year, 1970-1971. Volume I. X, 661 pages. 1972. DM 40,-

Vol. 247: Lectures on Operator Algebras. Tulane University Ring and Operator Theory Year, 1970-1971. Volume II. XI, 786 pages. 1972. DM 40,-

Vol. 248: Lectures on the Applications of Sheaves to Ring Theory. Tulane University Ring and Operator Theory Year, 1970-1971. Volume III. VIII, 315 pages. 1971. DM 26,-

Vol. 249: Symposium on Algebraic Topology. Edited by P. J. Hilton. VII, 111 pages. 1971. DM 16,-

Vol. 250: B. Jónsson, Topics in Universal Algebra. VI, 220 pages. 1972. DM 20,-

Vol. 251: The Theory of Arithmetic Functions. Edited by A. A. Gioia and D. L. Goldsmith VI, 287 pages. 1972. DM 24,-

Vol. 252: D. A. Stone, Stratified Polyhedra. IX, 193 pages. 1972. DM 18,-

Vol. 253: V. Komkov, Optimal Control Theory for the Damping of Vibrations of Simple Elastic Systems. V, 240 pages. 1972. DM 20,-

Vol. 254: C. U. Jensen, Les Foncteurs Dérivés de $\varprojlim$ et leurs Applications en Théorie des Modules. V, 103 pages. 1972. DM 16,-

Vol. 255: Conference in Mathematical Logic - London '70. Edited by W. Hodges. VIII, 351 pages. 1972. DM 26,-

Vol. 256: C. A. Berenstein and M. A. Dostal, Analytically Uniform Spaces and their Applications to Convolution Equations. VII, 130 pages. 1972. DM 16,-

Vol. 257: R. B. Holmes, A Course on Optimization and Best Approximation. VIII, 233 pages. 1972. DM 20,-

Vol. 258: Séminaire de Probabilités VI. Edited by P. A. Meyer. VI, 253 pages. 1972. DM 22,-

Vol. 259: N. Moulis, Structures de Fredholm sur les Variétés Hilbertiennes. V, 123 pages. 1972. DM 16,-

Vol. 260: R. Godement and H. Jacquet, Zeta Functions of Simple Algebras. IX, 188 pages. 1972. DM 18,-

Vol. 261: A. Guichardet, Symmetric Hilbert Spaces and Related Topics. V, 197 pages. 1972. DM 18,-

Vol. 262: H. G. Zimmer, Computational Problems, Methods, and Results in Algebraic Number Theory. V, 103 pages. 1972. DM 16,-

Vol. 263: T. Parthasarathy, Selection Theorems and their Applications. VII, 101 pages. 1972. DM 16,-

Vol. 264: W. Messing, The Crystals Associated to Barsotti-Tate Groups: With Applications to Abelian Schemes. III, 190 pages. 1972. DM 18,-

Vol. 265: N. Saavedra Rivano, Catégories Tannakiennes. II, 418 pages. 1972. DM 26,-

Vol. 266: Conference on Harmonic Analysis. Edited by D. Gulick and R. L. Lipsman. VI, 323 pages. 1972. DM 24,-

Vol. 267: Numerische Lösung nichtlinearer partieller Differential- und Integro-Differentialgleichungen. Herausgegeben von R. Ansorge und W. Törnig, VI, 339 Seiten. 1972. DM 26,-

Vol. 268: C. G. Simader, On Dirichlet's Boundary Value Problem. IV, 238 pages. 1972. DM 20,-

Vol. 269: Théorie des Topos et Cohomologie Etale des Schémas. (SGA 4). Dirigé par M. Artin, A. Grothendieck et J. L. Verdier. XIX, 525 pages. 1972. DM 50,-

Vol. 270: Théorie des Topos et Cohomologie Etale des Schémas. Tome 2. (SGA 4). Dirigé par M. Artin, A. Grothendieck et J. L. Verdier. V, 418 pages. 1972. DM 50,-

Vol. 271: J. P. May, The Geometry of Iterated Loop Spaces. IX, 175 pages. 1972. DM 18,-

Vol. 272: K. R. Parthasarathy and K. Schmidt, Positive Definite Kernels, Continuous Tensor Products, and Central Limit Theorems of Probability Theory. VI, 107 pages. 1972. DM 16,-

Vol. 273: U. Seip, Kompakt erzeugte Vektorräume und Analysis. IX, 119 Seiten. 1972. DM 16,-

Vol. 274: Toposes, Algebraic Geometry and Logic. Edited by. F. W. Lawvere. VI, 189 pages. 1972. DM 18,-

Vol. 275: Séminaire Pierre Lelong (Analyse) Année 1970-1971. VI, 181 pages. 1972. DM 18,-

Vol. 276: A. Borel, Représentations de Groupes Localement Compacts. V, 98 pages. 1972. DM 16,-

continuation on page 185

Lecture Notes in Mathematics

Edited by A. Dold and B. Eckmann

435

Charles F. Dunkl
Donald E. Ramirez

Representations of Commutative Semitopological Semigroups

Springer-Verlag
Berlin · Heidelberg · New York 1975

Dr. Charles F. Dunkl
Dr. Donald E. Ramirez
Dept. of Mathematics
University of Virginia
Charlottesville, VA 22903/USA

Library of Congress Cataloging in Publication Data

Dunkl, Charles F 1941-
Representations of commutative semitopological semigroups.

(Lecture notes in mathematics ; 435)
Bibliography: p.
Includes indexes.
1. Topological groups. 2. Semigroups. 3. Representations of groups. I. Ramirez, Donald E., joint author. II. Title. III. Series: Lecture notes in mathematics (Berlin) ; 435.
QA3.L28 no. 435 [QA387] 510'.8s [512'.55] 74-32450

AMS Subject Classifications (1970): 22A25, 22B99, 43A30, 43A65, 46C10, 46H15

ISBN 978-3-540-07132-7 Springer-Verlag Berlin · Heidelberg · New York
ISBN 978-0-387-07132-9 Springer-Verlag New York · Heidelberg · Berlin

Offsetdruck: Julius Beltz, Hemsbach/Bergstr.

Preface

Harmonic analysis is primarily the study of functions and measures on topological spaces which also have an algebraic structure. In this book, the underlying structure is given by a commutative associative separately continuous multiplication, that is, a commutative semitopological semigroup (CSS). Everybody knows that representation theory is useful in studying almost all (maybe even all) mathematical structures. The theory of representing CSS's in compact CSS's is essentially the same as the theory of weakly almost periodic functions (see Eberlein [1], de Leeuw and Glicksberg [1], Berglund and Hofmann [1]).

To discover more structure, we investigate representations of CSS's in objects native to harmonic analysis. In order of increasing generality, they are the unit disc, the unit ball in an L^∞-space, the unit ball of the quotient of a function algebra, and the unit ball in the algebra of bounded operators on a Hilbert space. The latter three are furnished with weak topologies in which multiplication is separately continuous and the unit balls are compact.

This point of view provides a unified framework for diverse ideas like semicharacters, positive-definite functions, completely monotone functions, the Hausdorff moment problem, functions of bounded variation, the Plancherel theorem, duality for locally compact abelian groups, representation of uniquely divisible CSS's (the theory of Brown and Friedberg [1]), dilation theory on Hilbert space, et cetera. Interesting new problems can be posed even for such seemingly trivial semigroups as $X \cup \{0\}$, X any set, $xy = 0$, $x0 = 0$ $(x,y \in X)$.

We hope that this book will be interesting to both commutative harmonic analysis and topological semigroup people. As Berglund and Hofmann [1] point out, functional analysis is important in studying semitopological, rather than topological, semigroups, so we expect the reader to be acquainted with basic functional analysis (Rudin's book [2], for example). Some knowledge of duality theory for locally compact abelian groups would be helpful.

A substantial part of the work is new. Unless otherwise stated, semigroups will be assumed to be commutative, semitopological, and written multiplicatively. The usual exceptions to the latter are Z, Z_+, $\mathbb{R}$, $\mathbb{R}_+$ (integers, nonnegative integers, reals, nonnegative reals respectively) and their Cartesian products, which will always have the additive structure. Indices for authors, subjects, and symbols are provided.

We thank Marie Brown for her skillful typing of this manuscript.

During the preparation of this monograph, the authors were partially supported by NSF Grant GP-31483X.

C.F.D.

D.E.R.

Table of Contents

1.1.3

Chapter 1. Basic Results

In this chapter we collect the basic results about commutative semitopological semigroups and commutative C* and W* algebras which we will need in our investigation of CSS's. The reader familiar with these topics can simply skip to Chapter 2.

§1. Commutative semitopological semigroups

1.1 Definition: A semitopological semigroup S is a semigroup S with a Hausdorff topology such that the map $(x,y) \mapsto xy : S \times S \to S$ is separately continuous. A right ideal I in S satisfies $IS \subset I$. An ideal will mean a two-sided ideal.

The basic theory of compact semitopological semigroups has been discussed by Berglund and Hofmann [1] and Burckel [1]. We quote here some of the fundamental results of the theory of semitopological semigroups.

1.2 Proposition: *Let S be a semitopological semigroup and T a subsemigroup of S. Then $\overline{T}$ is a subsemigroup. If T is commutative, so is $\overline{T}$.*

1.3 Theorem (Ellis [1]): *Let X be a locally compact Hausdorff space with a group structure such that the map $(x,y) \mapsto xy$: $X \times X \to X$ is separately continuous. Then X is a topological group.*

1.4 Remark: For X a compact space and as in Theorem 1.3, deLeeuw and Glicksberg [1] have given a proof using invariant means (see also Burckel [1, p. 17]). Lawson [1] has studied extensions of Ellis' theorem. One of Lawson's results is that any subgroup G of a compact semitopological semigroup S is a topological group. We note that if S_μ is the unit ball of $L^\infty(\mu)$ (μ a probability measure) with the weak-* topology $\sigma(L^\infty(\mu), L^1(\mu))$ and with pointwise multiplication, then any subgroup G of S_μ is topological and metrizable. Indeed, the topology on G is the norm topology on G as a subset of $L^1(\mu)$: for let $f \in G$ and $\{f_\alpha\} \subset G$ with $f_\alpha \overset{\alpha}{\to} f$ weak-* in S_μ, then $|f_\alpha| = |f|$ and so

$$\{\int |f_\alpha - f| d\mu\}^2 \leq \int |f_\alpha - f|^2 \, d\mu$$

$$= \int |f_\alpha|^2 \, d\mu - \int f \, \overline{f}_\alpha d\mu - \int f_\alpha \overline{f} \, d\mu + \int |f|^2 \, d\mu$$

$$\overset{\alpha}{\to} \int |f|^2 \, d\mu - \int f \, \overline{f} \, d\mu - \int f \, \overline{f} \, d\mu + \int |f|^2 \, d\mu \quad = 0.$$

1.5 <u>Definition</u>: The <u>kernel</u> of a semigroup S is the intersection of all the nonempty two-sided ideals of S, and it is denoted by K(S).

1.6 <u>Corollary</u>: <u>Let</u> S <u>be a compact commutative semitopological semigroup. Then</u> K(S) <u>is the intersection of all the nonempty closed ideals of</u> S, <u>and</u> K(S) <u>is a nonempty compact abelian group</u>.

Proof. Let K be the intersection of all the closed (two-sided) ideals of S. Thus, $K \neq \emptyset$ (by the finite-intersection property for compact spaces) and is the minimal closed ideal in S. Let I be any ideal in S. Then for $x \in I$, $xS \subset I$; and since xS is a closed ideal K(S) = K. By the minimality of K(S), xK(S) = K(S) for $x \in S$.

1.1.8

Let $x \in K(S)$; there exists $e \in K(S)$ with $ex = x$. Since $xK(S) = K(S)$, for any $y \in K(S)$ we have $ey = ext$ (some $t \in K(S)$ with $xt = y$) $= xt = y$. So e is an identity for $K(S)$. Let $x' \in K(S)$ with $xx' = e$. Suppose $x'' \in K(S)$ has $xx'' = e$. Then $x' = ex' = xx''x' = ex'' = x''$, so x' is unique. Thus $K(S)$ is a group. Now apply (Ellis') Theorem 1.3. □

1.7 <u>Remark</u>: deLeeuw and Glicksberg [1] have shown for any compact semitopological semigroup S that $K(S) \neq \emptyset$ (see Burckel [1, p. 19]). The reader is warned that K(S) in general need not a group. For example, take $S = \{0,1\}$ with the left zero multiplication: $0 \cdot 0 = 0$, $0 \cdot 1 = 0$, $1 \cdot 1 = 1$, $1 \cdot 0 = 1$. Then $K(S) = S$. Also K(S) need not be closed (see Berglund and Hofmann [1, p. 156]).

1.8 <u>Corollary</u>: <u>A cancellative commutative compact semitopological semigroup</u> S <u>is a group</u>.

Proof. It suffices to show $K(S) = S$. Let $x \in S$ and choose $y \in S$ with $xy = e$, e the idempotent of $K(S)$. Thus $exy = e^2 = e = xy$ which implies by cancellation $ex = x$. Thus $x \in eS \subset K(S)$. □

1.9 Definition: A semicharacter ϕ of a semitopological semigroup S is a nonzero continuous homomorphism of S into the unit disk of the complex numbers (under multiplication). The set of all semicharacters is denoted by $\hat{S}$.

1.10 Definition: Given an idempotent ε in a commutative semigroup S, define $H(\varepsilon) = \{x \in S: \varepsilon x = x$ and there exists $y \in S$ with $xy = \varepsilon\}$. The set $H(\varepsilon)$ is the maximal group about ε.

1.11 Remark: One of the important differences between topological (jointly continuous) semigroups and semitopological semigroups (both commutative) is that although $H(\varepsilon)$ is closed for topological semigroups (ε an idempotent), this need not occur in semitopological semigroups.

1.12 Remark: For S a compact semitopological semigroup, if $\hat{S}$ separates the points of S then S is topological: $x_\alpha \overset{\alpha}{\to} x$ in S if and only if $\phi(x_\alpha) \overset{\alpha}{\to} \phi(x)$ for all $\phi \in \hat{S}$ (x_α, $x \in S$) and note $x_\alpha \overset{\alpha}{\to} x$ and $y_\beta \overset{\beta}{\to} y$ implies $\phi(x_\alpha y_\beta) = \phi(x_\alpha)\phi(y_\beta) \xrightarrow{\alpha,\beta} \phi(x)\phi(y)$ $= \phi(xy)$, $(x, x_\alpha, y, y_\beta \in S, \phi \in \hat{S})$. Thus S can be embedded into a product of discs.

§2. C* and W*-algebras

In this section we collect some of the fundamental facts concerning C* and W*-algebras. Our basic reference will be Sakai [1].

2.1 Definition: A *-algebra A is a complex algebra with an involution $x \mapsto x^*: A \to A$ such that $(x^*)^* = x$, $(x+y)^* = x^*+y^*$, $(xy)^* = y^*x^*$, and $(\lambda x)^* = \bar{\lambda}x^*$ $(x \in A, \lambda \in \mathbb{C})$.

If A is a Banach $*$-algebra with $||x^*x|| = ||x||^2$ $(x \in A)$, then A is called a C*-algebra.

If, in addition, A is a dual space, then A is called a W*-algebra.

2.2 Examples (1): The space $C^B(X)$ of bounded continuous functions on a locally compact space X is a commutative C*-algebra.

(2): The space $\mathcal{B}(\mathcal{H})$ of bounded operators on the Hilbert space $\mathcal{H}$ is a W*-algebra. The predual of $\mathcal{B}(\mathcal{H})$ is the Banach space of all trace class operators on $\mathcal{H}$, (Sakai [1, p. 39]).

2.3 Theorem (Gelfand-Naimark [1]): A commutative C*-algebra A with unit is isometrically $*$-isomorphic to the algebra of all complex continuous functions on its maximal ideal space $\mathcal{M}_A$; that is $A \cong C(\mathcal{M}_A)$. If A does not possess a unit, then $A \cong C_o(\mathcal{M}_A)$, the algebra of all complex continuous functions on $\mathcal{M}_A$ which vanish at infinity.

Proof. See Sakai [1, p. 4]. □

2.4 Theorem: W*-algebras always possess a unit.

Proof. The unit ball of a C*-algebra A has extreme points if and only if A has an identity (Sakai [1, p. 10]). But since a W*-algebra is a dual space, its unit ball has plenty of extreme points (by the Krein-Milman theorem).

2.5 Definition: Let x be an element in a Banach algebra A with unit e. The spectrum $\sigma(x)$ of x consists of those complex numbers λ such that $\lambda e - x$ is singular (has no inverse in A). (If A has no unit, then we adjoin one.) The spectral radius $||x||_{sp}$ of $x \in A$ is $\sup \{|\lambda| : \lambda \in \sigma(x)\}$. The spectral radius has the property

$||x||_{sp} = \lim_{n\to\infty} ||x^n||^{1/n} \leq ||x||_A \quad (x \in A)$. For a commutative C*-algebra A, $||x||_{sp} = ||x||_A \quad (x \in A)$.

2.6 Kaplansky Density Theorem (Kaplansky [1]): Let A be a self-adjoint subalgebra of the W*-algebra B. Let B_* be the predual of B. If A is $\sigma(B,B_*)$-dense in B, then the unit ball of A is $\sigma(B,B_*)$-dense in the unit ball of B.

Proof. See Sakai [1, p. 20]. □

2.7 Theorem: Every C*-algebra is *-isomorphic to a norm closed self-adjoint subalgebra of $\mathcal{B}(\mathcal{H})$, for some Hilbert space $\mathcal{H}$. Every W*-algebra is *-isomorphic to a weak-operator closed self-adjoint subalgebra with unit of $\mathcal{B}(\mathcal{H})$, some Hilbert space $\mathcal{H}$.

Proof. See Sakai [1, p. 41]. □

2.8 Definition: A state ϕ on the W*-algebra A is a bounded linear functional with $||\phi|| = 1$ and $\phi(x^*x) \geq 0 \quad (x \in A)$. A normal state is one which is $\sigma(A,A_*)$-continuous. This is equivalent to the condition: $\phi(\sup_\alpha x_\alpha) = \sup_\alpha \phi(x_\alpha)$ for every uniformly bounded increasing net $\{x_\alpha\} \subset A$, (Sakai [1, p. 28]).

2.9 Remark: If A is a commutative W*-algebra (so $A \cong C(M_A)$), then each normal state ϕ defines a positive measure (called "normal") $\mu_\phi \in M(M_A)$. The map of A onto $L^\infty(\mu_\phi)$ is a *-homomorphism. Further, the map of A into $\Sigma_\phi \oplus L^\infty(\mu_\phi)$ (the direct sum over all states) is a *-isomorphism. The map is onto $\Sigma_\phi \oplus L^\infty(\mu_\phi)$ provided the direct sum is taken over a certain family Φ of normal states (Sakai [1, p. 45-46]).

2.10 Remark: Let A be a commutative W*-algebra, with $A \cong \Sigma_\phi \oplus L^\infty(\mu_\phi)$ $(\phi \in \Phi)$; and we may write $A \cong L^\infty(\Gamma,\nu)$ where $\Gamma = \bigcup_\phi \operatorname{spt}\mu_\phi$ ($\operatorname{spt}\mu_\phi$ is the closed support of μ_ϕ) and $\nu = \Sigma_\phi \oplus \mu_\phi$; thus (Γ,ν) is a direct sum of finite measure spaces (called a localizable space). Let $L^1(\Gamma,\nu)$ be the Banach space of all ν-integrable functions on Γ, then A_* (the predual of A) is isomorphic to the complex L-space $L^1(\Gamma,\nu)$.

2.11 Example: Let X be a locally compact Hausdorff space and let M(X) be the Banach space of all finite regular Borel measures on X. Then M(X)* is a commutative W*-algebra:

For $F \in M(X)^*$ and $\mu \in M(X)$, write $\langle\mu,F\rangle$ for the canonical pairing of M(X) with its dual space M(X)*. For each $\mu \in M_p(X)$ (the space of probability measures) let $F_\mu = F|L^1(\mu)$ $(F \in M(X)^*)$. Thus $F_\mu \in L^1(\mu)^* = L^\infty(\mu)$ and $||F_\mu||_\infty \le ||F||$. Indeed, $||F|| = \sup\{||F_\mu||_\infty : \mu \in M_p(X)\}$. Define for $F \in M(X)^*$, F^* by $(F^*)_\mu = (F_\mu)^-$ $(\mu \in M_p(X))$. For $F \in M(X)^*$ and $\mu \in M(X)$, define $F\cdot\mu \in M(X)$ by $\int_X h \, d(F\cdot\mu) = \int_X h F_\mu \, d\mu$ $(h \in C_o(X))$. Thus M(X) is an M(X)*-module. For $F,G \in M(X)^*$ the multiplication in M(X)* is defined by $\langle\mu,FG\rangle = \langle F\cdot\mu,G\rangle$ $(\mu \in M(X))$. Thus $(FG)_\mu = F_\mu G_\mu$ and $||FG|| \le ||F|| \, ||G||$ $(F,G \in M(X)^*, \mu \in M(X))$. That M(X)* is a commutative Banach $*$-algebra is straightforward. To see that it is a C*-algebra note that $||FF^*|| =$ $\sup\{||(FF^*)_\mu||_\infty : \mu \in M_p(X)\} = \sup\{||F_\mu \overline{F}_\mu||_\infty : \mu \in M_p(X)\}$ $= (\sup\{||F_\mu||_\infty : \mu \in M_p(X)\})^2 = ||F||^2$. Since M(X)* is a dual space, it is a W*-algebra. Further M(X)* is a projective limit of $\{L^\infty(\mu) : \mu \in M_p(X)\}$ since for $\mu_1 << \mu_2$ (μ_1 absolutely continuous with respect to μ_2) $F_{\mu_1} = F_{\mu_2}$ (μ_2 - a.e.) $(\mu_1,\mu_2 \in M_p(X))$. In general, the second dual of a C*-algebra is W*, (Sakai [1, p. 43]).

2.12 Definition: A compact Hausdorff space X is called Stonean (or extremally disconnected) if and only if the closure of every open set is open.

2.13 Remark: If a compact space X is such that every bounded increasing net of real-valued, nonnegative functions in C(X) has a least upper bound in C(X), then X is Stonean (Sakai [1, p. 6]). Since W*-algebras possess the property that a uniformly bounded, increasing net of elements converges to its least upper bound (Sakai [1, p. 15]), one has that the maximal ideal space of a commutative W*-algebra is a Stonean space. Given X compact Hausdorff, for C(X) to be a W*-algebra it is necessary and sufficient that X is hyper-Stonean; i.e. X is Stonean and has a faithful family $\{\mu_\alpha : \alpha \in \Phi\}$ of normal measures (if $f \in C(X)$ with $f \geq 0$ and $\int_X f\, d\mu_\alpha \geq 0$ for all $\alpha \in \Phi$, then $f = 0$), (Dixmier [1], Sakai [1, p. 46]).

2.14 Remark: For a locally compact Hausdorff space X, M(X) is weakly sequentially complete (that is, every weakly Cauchy sequence converges) (Dunford and Schwartz [1, p. 311]). The general result is that the predual of a W*-algebra is weakly sequentially complete (Sakai [2]). On the other hand, a closed subalgebra of a commutative C*-algebra is never weakly sequentially complete unless finite dimensional (see 2.16).

2.15 Definition: A function algebra A is a closed subalgebra of $C_o(X)$ (X a locally compact Hausdorff space) which separates the points of X in the sense that if $x,y \in X$, $x \neq y$, then there exists $f \in A$ with $f(x) \neq f(y)$.

2.16 Theorem: *If a function algebra A is weakly sequentially complete then A is finite dimensional.*

Proof. If A is infinite dimensional, then A has an infinite dimensional separable subalgebra which is also weakly sequentially complete. Thus we assume A is separable.

The algebra $A \subset C_o(X)$ separates the points in X and hence in the Silov boundary ∂A (since $\partial A \subset X$). Thus ∂A is a metrizable locally compact space. Let $P \subset \partial A \subset X$ denote the set of peak points of A. The set P is dense in ∂A (by Bishop's peak point theorem (Gamelin [1, p. 56])) since ∂A is metrizable. It will thus suffice to show P is a finite set: for then ∂A will be a finite set (and equal to P), and A is isomorphic to $A|\partial A$.

By the Lebesgue dominated convergence theorem, given a sequence $\{f_n\} \subset A$ with $f_n \overset{n}{\to} \chi_p$ (the characteristic function of the set $\{p\}$, $p \in P$) pointwise on X, it follows that $\{f_n\}$ is weakly Cauchy in A. Since A is weakly sequentially complete, $\chi_p \in A$. Thus P consists of isolated points.

By the weak sequential completeness of A and the Lebesgue dominated convergence theorem, P is finite - for otherwise, we would have a countable subset $Q \subset P$ with $\chi_Q \in A \subset C_o(X)$. But Q would then be a compact infinite discrete set, a contradiction.□

2.17 Remark: That $C_o(X)$ is weakly sequentially complete if and only if X is finite (X locally compact Hausdorff) is used by R. Edwards [1] to show that the Fourier-Stieltjes transforms of the measure algebra M(G) of a locally compact abelian group G is never onto unless G is finite. Theorem 2.16 appears in Dunkl and Ramirez [3].

Chapter 2. The Representation Algebra

§1. L^∞-representations

The starting point of our investigations is the fact that the unit ball in L^∞ of a probability space is a compact commutative semitopological semigroup under multiplication and the weak-* topology.

Let (μ,Ω) be a probability space, that is, μ is a positive measure on the measure space Ω with $\mu\Omega = 1$. Then $L^1(\mu,\Omega)$ denotes the Banach space of (equivalence classes) of μ-integrable functions on Ω, with $||g||_1 = \int_\Omega |g| d\mu$, $g \in L^1(\mu,\Omega)$; and $L^\infty(\mu,\Omega)$ is the commutative W*-algebra of essentially bounded measurable functions on Ω, with norm $||h||_\infty = \text{essup}_\Omega |h|$, $h \in L^\infty(\mu,\Omega)$. The dual space of $L^1(\mu,\Omega)$ is $L^\infty(\mu,\Omega)$, so the <u>weak-* topology</u> on $L^\infty(\mu,\Omega)$ is given by: a net $\{f_\alpha\} \subset L^\infty(\mu,\Omega)$ converges weak-* to $f \in L^\infty(\mu,\Omega)$ if and only if $\int_\Omega f_\alpha g d\mu \overset{\alpha}{\to} \int_\Omega fg d\mu$ for each $g \in L^1(\mu,\Omega)$. The closed unit ball $\{f \in L^\infty(\mu,\Omega): ||f||_\infty \leq 1\}$ is closed under multiplication since $||f_1 f_2||_\infty \leq ||f_1||_\infty ||f_2||_\infty$ $(f_1, f_2 \in L^\infty(\mu,\Omega))$ and is weak-* compact (Alaoglu's theorem, (Dunford and Schwartz [1, p. 424])). Note also that $L^\infty(\mu,\Omega)$ acts as an algebra of bounded operators on the Hilbert space $L^2(\mu,\Omega)$, and the weak-* topology is identical to the <u>weak-operator</u> (WO) <u>topology</u> (the net $\{f_\alpha\} \subset L^\infty(\mu,\Omega)$ converges WO to $f \in L^\infty(\mu,\Omega)$ if and only if $\int_\Omega (f_\alpha g_1)\overline{g}_2 d\mu \overset{\alpha}{\to} \int_\Omega (fg_1)\overline{g}_2 d\mu$ for all $g_1, g_2 \in L^2(\mu,\Omega)$).

2.1.1

By way of motivation, note that the semigroup [0,1] with the "min" operation is commutative, compact, and topological but has no continuous semicharacters other than the constants 1 and 0. However mapping $x \in [0,1]$ onto the characteristic function $\chi_{[0,x]}$ of the interval $[0,x]$, we obtain a weak-* continuous representation of the semigroup in $L^\infty(\mu,[0,1])$, where μ is any continuous (null on countable sets) probability measure μ on [0,1], (see Section 2 for more details). This representation is faithful (one-to-one) if μ is supported on all of [0,1]. We will now further develop this theory, which is richer than that of continuous semicharacters.

In what follows S will denote an infinite commutative semitopological semigroup, $C^B(S)$ denotes the space of bounded continuous functions on S, and f_y is the <u>translate</u> of $f \in C^B(S)$ by y, that is, $f_y(x) = f(xy)$, $(x,y \in S)$.

1.1 <u>Definition</u>: An L^∞-<u>representation</u> (T,μ,Ω) of S is a weak-* continuous homomorphism T of S into the unit ball of $L^\infty(\mu,\Omega)$. Thus T has the following properties:

a) for each $x \in S$, $Tx \in L^\infty(\mu,\Omega)$ and $||Tx||_\infty \leq 1$;

b) $x_\alpha \overset{\alpha}{\to} x$ in S implies $Tx_\alpha \overset{\alpha}{\to} Tx$ weak-*;

c) $x,y \in S$ implies $T(xy) = (Tx)(Ty)$ a.e. .

We note that an L^∞-representation gives a homomorphism of S into a dense subsemigroup of a compact semitopological semigroup (namely, the weak-* closure of $\{Tx : x \in S\}$).

2.1.2

At first glance, the class of all L^∞-representations of a given semigroup, although interesting, seems to be unmanageably large, for example if $S = Z_+ = \{0,1,2,\cdots\}$ under addition, then any probability space (μ,Ω) with any function $h \in L^\infty(\mu,\Omega)$ with $||h||_\infty \leq 1$ will give a representation $Tn = h^n$, $n \in Z_+$. However we can simplify the situation to where it suffices to consider a certain algebra of continuous functions on S, which we will now define.

1.2 <u>Definition</u>: The <u>representation algebra</u>[1] of S denoted by $\mathcal{R}(S)$, is the set of functions $f(x) = \int_\Omega (Tx) g d\mu$, $(x \in S)$, where (T,μ,Ω) is an L^∞-representation of S, and $g \in L^1(\mu,\Omega)$.

Note that $\mathcal{R}(S) \subset C^B(S)$ since $|f(x)| \leq ||g||_1$ and $x_\alpha \overset{\alpha}{\to} x$ in S implies $Tx_\alpha \overset{\alpha}{\to} Tx$ weak-*. For convenience, we will sometimes write $T^*g(x) = \int_\Omega (Tx) g d\mu$, $g \in L^1(\mu,\Omega)$.

We will show that $\mathcal{R}(S)$ is a Banach algebra which is proper in $C^B(S)$ (or indeed any infinite-dimensional C*-algebra), but first we must clear up a set-theoretical paradox. We cannot form the set of all representations, but we can form a set of L^∞-representations of S which suffices. Let θ be the cardinality of S. Given an L^∞-representation (T,μ,Ω) we may assume that Ω is a compact Stonean space, and $L^\infty(\mu,\Omega)$ is identified with $C(\Omega)$. The closed algebra generated by $\{Tx : x \in S\} \cup \{\overline{Tx} : x \in S\}$ defines an identification space Ω_1 of Ω which can have at most c^θ points (c = cardinality of the reals). The natural injection $C(\Omega_1) \to C(\Omega)$ induces an adjoint map which takes μ to a

[1] The term representation algebra introduced here differs from that used by Hofmann [1, p. 89] who gives a duality theory for compact topological semigroups.

probability measure μ_1 on Ω_1 and $L^1(\mu,\Omega)$ onto $L^1(\mu_1,\Omega_1)$. Also for each $g \in L^1(\mu,\Omega)$ there exists $g_1 \in L^1(\mu_1,\Omega_1)$ such that $\int_\Omega (Tx)\, g d\mu = \int_{\Omega_1} (Tx)\, g_1 d\mu_1$ and $||g_1||_1 \leq ||g||_1$. (Note that $Tx \in C(\Omega_1)$ so (T,μ_1,Ω_1) is also an L^∞-representation.) We can now make the following definition:

1.3 <u>Definition</u>: Let S denote the set of all L^∞-representations (T,μ,Ω) of S such that the cardinality of Ω is less than or equal to $(\text{card } \mathbb{R})^{\text{card } S}$. (Note that Definition 1.2 can (and should) be given in terms of S alone.)

1.4 <u>Proposition</u>: $R(S) \subset WAP(S)$.

Proof. Recall WAP(S) is the space of $f \in C^B(S)$ such that $\{f_y : y \in S\}$ is weakly relatively compact (Burckel [1, p. 1]). Let $f \in R(S)$, then there exists an L^∞-representation (T,μ,Ω) and $g \in L^1(\mu,\Omega)$ such that $f = T^*g$. Note that $f_y(x) = \int_\Omega T(xy)\, g d\mu = \int_\Omega (TxTy)\, g d\mu = T^*((Ty)g)(x)$ and that T^* is a continuous linear map: $L^1(\mu,\Omega) \to C^B(S)$. But the set $E = \{hg : h \in \text{weak-* closure } \{Tx : x \in S\}\}$ is weakly compact in $L^1(\mu,\Omega)$, and T^* is weakly continuous (Dunford and Schwartz [1, p. 422]). Thus $\{f_y : y \in S\} \subset T^*E$ which is weakly compact in $C^B(S)$. □

We will show that $R(S)$ is significantly different from WAP(S), although it is a Banach space of continuous functions.

1.5 <u>Definition</u>: For $f \in R(S)$, define the norm of f to be
$$||f||_R = \inf\{||g||_1 : (T,\mu,\Omega) \in S,\ g \in L^1(\mu,\Omega),\ f(x) = \int_\Omega (Tx)\, g d\mu\}.$$

Note that the sup-norm of f, $||f||_\infty \le ||f||_R$.

The above remarks show that this infimum would not change if S were enlarged. We will now show that $R(S)$ is a Banach algebra. The device will be that of forming direct sums and tensor products of L^∞-representations.

1.6 Theorem: $R(S)$ is a normed algebra, under the pointwise operations on S. It contains the constant functions, and is closed under conjugation. If $f \in R(S)$, $y \in S$ then $||f_y||_R \le ||f||_R$.

Proof. 1) From Definition 1.5, we see that $||af||_R = |a|\ ||f||_R$ ($a \in \mathbb{C}$, $f \in R(S)$) and that $||f||_R = 0$ implies $||f||_\infty = 0$ and thus $f = 0$.

2) Let $f_1, f_2 \in R(S)$. We will show that f_1+f_2 and $f_1 f_2 \in R(S)$ and that $||f_1+f_2||_R \le ||f_1||_R + ||f_2||_R$ and $||f_1 f_2||_R \le ||f_1||_R ||f_2||_R$. Given $\varepsilon > 0$, there exist $(T_i, \mu_i, \Omega_i) \in S$, $g_i \in L^1(\mu_i, \Omega_i)$ such that $f_i = T_i{}^* g_i$ and $||g_i||_1 < ||f_i||_R + \varepsilon/2$ (for $i = 1,2$). We may assume Ω_1 and Ω_2 are disjoint.

2a) We form the direct sum representation of T_1 and T_2. Let $\Omega = \Omega_1 \cup \Omega_2$ and define the probability measure μ on Ω to be $(1/2)\mu_i$ on Ω_i $(i = 1,2)$. For $x \in S$, define $Tx \in L^\infty(\mu,\Omega)$ to be $T_i x$ on Ω_i $(i = 1,2)$. It is easy to check that (T,μ,Ω) is an L^∞-representation. Now let $g = 2g_i$ on Ω_i $(i = 1,2)$, $x \in S$, then $T^*g(x) = \int_\Omega (Tx)g\,d\mu = \int_{\Omega_1}(T_1x)g_1 d\mu_1 + \int_{\Omega_2}(T_2x)g_2 d\mu_2$ $= f_1(x) + f_2(x)$, so $f_1 + f_2 \in R(S)$, and

2.1.6

$||f_1+f_2||_R \le ||g||_1 = ||g_1||_1 + ||g_2||_1 < ||f_1||_R + ||f_2||_R + \varepsilon$, but ε was arbitrary > 0.

2b) We form the tensor product of T_1 and T_2. Let $\Omega = \Omega_1 \times \Omega_2$, and define $\mu = \mu_1 \times \mu_2$, a probability measure on Ω. For $x \in S$, put $Tx(\omega_1,\omega_2) = T_1x(\omega_1)T_2x(\omega_2)$ $(\omega_i \in \Omega_i,\ i = 1,2)$. Note that T is still a semigroup homomorphism, but we must check the weak-* continuity. The algebraic tensor product $L^1(\mu_1,\Omega_1) \otimes L^1(\mu_2,\Omega_2)$ is norm-dense in $L^1(\mu,\Omega)$ (by Fubini's theorem), so for $h \in L^1(\mu,\Omega)$ and $\delta > 0$ there exist $h_{ik} \in L^1(\mu_i,\Omega_i)$ $(i = 1,2;\ k = 1,2,\cdots,n)$ such that

$$\int_{\Omega_1}\int_{\Omega_2} |h(\omega_1,\omega_2) - \sum_{k=1}^{n} h_{1k}(\omega_1)h_{2k}(\omega_2)|\,d\mu_2(\omega_2)\,d\mu_1(\omega_1) < \delta.$$

Let $\{x_\alpha\}$ be a net in S with $x_\alpha \overset{\alpha}{\to} x \in S$, then

$$|\int_\Omega (Tx_\alpha)h-(Tx)h\,d\mu| = |\int_\Omega\{(Tx_\alpha)h-(Tx_\alpha)\Sigma h_{1k}h_{2k} + (Tx_\alpha)\Sigma h_{1k}h_{2k} - (Tx)\Sigma h_{1k}h_{2k} + (Tx)\Sigma h_{1k}h_{2k} - (Tx)h\}d\mu| \le \int_\Omega |Tx_\alpha - Tx|\,|h-\Sigma h_{1k}h_{2k}|\,d\mu$$
$$+\Sigma_k |\int_{\Omega_1}(T_1x_\alpha)h_{1k}d\mu_1 \int_{\Omega_2}(T_2x_\alpha)h_{2k}d\mu_2 - \int_{\Omega_1}(T_1x)h_{1k}d\mu_1\int_{\Omega_2}(T_2x)h_{2k}d\mu_2|.$$

The second term tends to 0 as $x_\alpha \to x$, and the first term is bounded by 2δ. Thus T is weak-* continuous.

Finally, let $g(\omega_1,\omega_2) = g_1(\omega_1)g_2(\omega_2)$, $x \in S$ then $T^*g(x) = \int_\Omega (Tx)g\,d\mu = f_1(x)f_2(x)$, so $f_1f_2 \in R(S)$, and $||f_1f_2||_R \le ||g||_1 = ||g_1||_1||g_2||_1 < (||f_1||_R + \varepsilon/2)(||f_2||_R + \varepsilon/2)$. Thus $||f_1f_2||_R \le ||f_1||_R||f_2||_R$.

3) From the trivial representation $S \to 1$ in any $L^\infty(\mu,\Omega)$ we see that the function $1 \in R(S)$ and $||1||_R = 1$.

2.1.7

4) For each $(T,\mu,\Omega) \in \mathcal{S}$, there is a conjugate representation $(\overline{T},\mu,\Omega)$: merely map $x \in S$ to $\overline{Tx} \in L^\infty(\mu,\Omega)$ and observe conjugation in $L^\infty(\mu,\Omega)$ is weak-* continuous. For $g \in L^1(\mu,\Omega)$, we have $\overline{T^*g(x)} = \int_\Omega (\overline{Tx})\overline{g}d\mu = \overline{T}^*\overline{g}(x)$, thus $\overline{T^*g} \in \mathcal{R}(S)$ and $||\overline{T^*g}||_{\mathcal{R}} \leq ||\overline{g}||_1 = ||g||_1$. By symmetry, $||\overline{f}||_{\mathcal{R}} = ||f||_{\mathcal{R}}$.

5) Let $f \in \mathcal{R}(S)$, $\varepsilon > 0$, then there exist (T,μ,Ω) and $g \in L^1(\mu,\Omega)$ such that $f = T^*g$ and $||g||_1 < ||f||_{\mathcal{R}} + \varepsilon$. For $y \in S$ we have $f_y = T^*((Ty)g)$, and
$||f_y||_{\mathcal{R}} \leq ||(Ty)g||_1 \leq ||Ty||_\infty ||g||_1 \leq ||g||_1 \leq ||f||_{\mathcal{R}} + \varepsilon$. □

1.7 Proposition: $\mathcal{R}(S)$ contains the sums of series of continuous (bounded) semicharacters with summable coefficients.

Proof. Let $\{\phi_n\}_{n=1}^\infty$ be a sequence of continuous semicharacters, and let $f = \Sigma_{n=1}^\infty a_n\phi_n$, with $0 < \Sigma_{n=1}^\infty |a_n| < \infty$. Let Ω be the space $\{1,2,\cdots\}$ with the measure $\mu\{n\} = |a_n|/\Sigma_{j=1}^\infty |a_j|$, and define $Tx(n) = \phi_n(x)$ $(x \in S,\ n = 1,2,\cdots)$. It is clear that (T,μ,Ω) is an L^∞-representation of S, and that $f = T^*g$, where $g(n) = (\operatorname{sgn} a_n)\ (\Sigma_{j=1}^\infty |a_j|)$, ($\operatorname{sgn} z = z/|z|$, for $z \neq 0$ and $\operatorname{sgn} 0 = 0$, $z \in \mathbb{C}$). Thus $f \in \mathcal{R}(S)$ and $||f||_{\mathcal{R}} \leq \Sigma_{j=1}^\infty |a_j|$. □

1.8 Remark: If ϕ is a bounded continuous semicharacter, then $\phi \in \mathcal{R}(S)$ and $||\phi||_{\mathcal{R}} \leq 1$. If further S has an identity e, and $\phi \neq 0$, then $1 = \phi(e) = ||\phi||_\infty \leq ||\phi||_{\mathcal{R}} \leq 1$, so $||\phi||_{\mathcal{R}} = 1$.

1.9 Theorem: $\mathcal{R}(S)$ is complete, and thus is a commutative Banach algebra of continuous functions.

2.1.9

Proof. Let $\{f_n\}$ be a Cauchy sequence in $R(S)$. Extract a subsequence $\{f_{n_k}\}_{k=1}^{\infty}$ such that $||f_{n_k} - f_{n_{k+1}}||_R < 2^{-k-1}$, $k = 1,2,\cdots$. There exist a sequence $\{(T_k,\mu_k,\Omega_k)\}_{k=0}^{\infty} \subset S$ and functions $g_k \in L^1(\mu_k,\Omega_k)$, $k = 0,1,2,\cdots$ such that

1) $f_{n_1} = T_0^* g_0$,

2) $f_{n_{k+1}} - f_{n_k} = T_k^* g_k$ $(k = 1,2,\cdots)$, and

3) $||g_k||_1 < 2^{-k}$.

We now form the countable direct sum $\Sigma_{k=0}^{\infty} \oplus T_k$. Assume $\Omega_j \cap \Omega_k = \emptyset$ for $j \neq k$, let $\Omega = \bigcup_{k=0}^{\infty} \Omega_k$, define the probability measure μ on Ω to be $2^{-k-1}\mu_k$ on Ω_k, and define $Tx \in L^{\infty}(\mu,\Omega)$ to be $T_k x$ on Ω_k $(k = 0,1,2,\cdots)$. By an argument similar to that in Theorem 1.6 we show that T is weak-* continuous (we need only the fact that elements of $L^1(\mu,\Omega)$ supported by finitely many Ω_k are norm dense in $L^1(\mu,\Omega)$ and that $||Tx-Ty||_{\infty} \leq 2$).

For $k = 0,1,2,\cdots$ define $h_k \in L^1(\mu,\Omega)$ by $h_k = 0$ off Ω_k, $h_k = 2^{k+1} g_k$ on Ω_k. Then $||h_k||_1 = ||g_k||_1$ and $T^* h_k = T_k^* g_k$. But $\Sigma_{k=0}^{\infty} h_k$ is absolutely convergent, since $||h_k||_1 < 2^{-k}$ for $k = 1,2,\cdots$. Let $h = \Sigma_{k=0}^{\infty} h_k \in L^1(\mu,\Omega)$, and let $f = T^* h \in R(S)$. Observe $f_{n_k} = \Sigma_{j=0}^{k-1} T_j^* g_j = T^* \Sigma_{j=0}^{k-1} h_j$, so $||f - f_{n_k}||_R \leq ||\Sigma_{j=k}^{\infty} h_j||_1$ $\leq \Sigma_{j=k}^{\infty} ||h_j||_1 \to 0$ as $k \to \infty$. Finally for any $m = 1,2,\cdots$, $||f-f_m||_R \leq ||f-f_{n_k}||_R + ||f_{n_k} - f_m||_R \to 0$ as $m \to \infty$ (taking n_k large enough and $m \geq n_k$). □

1.10 <u>Definition</u>: Let W be the space of S-tuples f such that for $(T,\mu,\Omega) \in S$, $f_T \in L^\infty(\mu,\Omega)$ and $||f||_\infty = \sup ||f_T||_\infty < \infty$. This is a commutative W*-algebra under the coordinatewise operations. It is also called $\Sigma \oplus_{T\in S} L^\infty(\mu,\Omega)$, and is the dual of $L = \Sigma \oplus_{T\in S} L^1(\mu,\Omega)$, the elements of which are S-tuples g such that for $T \in S$, $g_T \in L^1(\mu,\Omega)$, and $||g||_1 = \Sigma_{T\in S}||g_T||_1 < \infty$. Thus $g_T \neq 0$ for at most countably many T. For $f \in W$, $g \in L$, define the canonical pairing $\langle f,g\rangle = \Sigma_{T\in S}\int_\Omega f_T g_T d\mu$. For $x \in S$, define $\rho x \in W$ by $(\rho x)_T = Tx$, $(T \in S)$.

1.11 <u>Proposition</u>: ρ <u>is a weak-* continuous representation of</u> S, <u>and if</u> $g \in L$ <u>then the function</u> $f(x) = \langle \rho x,g\rangle \in R(S)$, <u>with</u> $||f||_R \leq ||g||_1$.

Proof. Let $g \in L$, then we can write g as $(g_n)_{n=1}^\infty$ where $g_n \in L^1(\mu_n,\Omega_n)$ and $||g||_1 = \Sigma_{n=1}^\infty ||g_n||_1 < \infty$. Let $f_n = T_n^* g_n \in R(S)$, then $\Sigma_{n=1}^\infty f_n$ converges absolutely in $R(S)$ to $f \in R(S)$ (by completeness, Theorem 1.9), and $||f||_R \leq \Sigma_{n=1}^\infty ||g_n||_1 = ||g||_1$. Hence for $x \in S, \langle \rho x,g\rangle = \Sigma_{n=1}^\infty \int_{\Omega_n} (T_n x) g_n d\mu = \Sigma_{n=1}^\infty T_n^* g_n(x) = f(x)$. This implies that $x \mapsto \rho x$ is weak-* continuous. □

The following is clear from Definition 1.5 and Proposition 1.11.

1.12 <u>Theorem</u>: <u>Let</u> ρ^* <u>be the map</u> $L \to R(S)$ <u>defined by</u> $\rho^* g(x) = \langle \rho x,g\rangle$, $(x \in S)$, <u>then this map induces an isometric isomorphism of</u> $L/\ker \rho^*$ <u>onto</u> $R(S)$.

1.13 Theorem: The dual of $R(S)$, denoted by $A(S)$, is isometrically isomorphic to the weak-* closed span of $\{\rho x : x \in S\}$ in W, thus $A(S)$ is a function algebra. The multiplication in $A(S)$ restricted to point evaluations on S is the image of the semigroup multiplication.

Proof. From Banach space theory, the dual of $R(S)$ is isometrically isomorphic to the annihilator of ker ρ^* in W. But $g \in$ ker ρ^* if and only if $\langle \rho x, g\rangle = 0$ for all $x \in S$, if and only if g annihilates $\{\rho x : x \in S\}$. Thus the annihilator of ker ρ^* is the weak-* closed span of $\{\rho x : x \in S\}$, an algebra.

For $y \in S$, define $\phi_y \in A(S)$ by $\phi_y(f) = f(y)$ $(f \in R(S))$. Under the isomorphism, ϕ_y is mapped to $\rho y \in W$, and so $\phi_x \cdot \phi_y$ is mapped to $(\rho x)(\rho y) = \rho(xy)$, the image of ϕ_{xy}. □

1.14 Corollary: $R(S)$ is not equal to any infinite-dimensional C*-algebra of continuous functions on S, in particular, $R(S) \subsetneq WAP(S)$, if infinite-dimensional.

Proof. Suppose not, then $R(S) = A$ as sets, where A is a sub-C*-algebra of $C^B(S)$, thus $A \cong C(X)$, $(X = M_A)$. Since $||f||_\infty \leq ||f||_R$, the map $R(S) \to A$ is bounded, and by the closed graph theorem the norms on $R(S)$ and A are equivalent. Thus the dual of A, namely the space M(X) of finite regular Borel measures on X, is isomorphic to $A(S)$. However M(X) is weakly sequentially complete, and $A(S)$ cannot be, unless finite-dimensional, since it is a function algebra (1.2.16). This contradiction finishes the proof. □

1.15 Remark: If $A(S)$ is conjugate closed (for example if S has the union of the maximal groups about the idempotents dense), then $A(S)$ is a commutative W*-algebra equal to $C(\Omega)$, Ω hyper-Stonean, and so its predual $R(S)$ is an L-space (1.2.10).

§2. Examples

We now compute $R(S)$ for a number of semigroups S.

2.1: Let S be the compact interval [0,1] under the "min" operation. Thus S is an idempotent semigroup, also called a semilattice. The set of continuous functions of bounded variation on [0,1] is denoted by $(C \cap BV)[0,1]$. The algebra $R(S) = (C \cap BV)[0,1]$; thus $R(S)$ is not in general a dual space:

Let $f \in R(S)$ with $f(x) = \int_\Omega (Tx)g\,d\mu$, $(T,\mu,\Omega) \in \mathcal{S}$. Since S is an idempotent semigroup $Tx = \chi_{E_x}$, the characteristic function of the set E_x $(x \in S)$, E_x determined modulo μ-null sets. For $\varepsilon > 0$, let $f(x) = \int_{E_x} g\,d\mu$ where $||g||_1 \le ||f||_R + \varepsilon$.

Given the partition $0 = x_0 < x_1 < \cdots < x_n = 1$ of [0,1], one has

$$|f(0)| + \Sigma_{j=1}^n |f(x_j) - f(x_{j-1})| = |\int_{E_0} g\,d\mu| + \Sigma_{j=1}^n |\int_{E_{x_j} \setminus E_{x_{j-1}}} g\,d\mu|$$

$$\le \int_\Omega |g|\,d\mu = ||g||_1 \le ||f||_R + \varepsilon, \ (x < y \text{ implies } \mu(E_x \setminus E_y) = 0).$$

Then $f \in (C \cap BV)[0,1]$ and $||f||_{BV} \le ||f||_R$.

Conversely, let $f \in (C \cap BV)[0,1]$. There exist a continuous measure $\mu \in M_p([0,1])$ and $g \in L^1(\mu)$ such that

$$f(x) = f(0) + \int_0^x g\,d\mu, \ (x \in S).$$

It follows that $f \in R(S)$ and

2.2.2

$$||f||_R \leq |f(0)| + ||g||_1 = ||f||_{BV}\,.$$

If $(C \cap BV)[0,1]$ were a dual space then the set $P = \{f \in (C \cap BV)[0,1] : f \geq 0,\ f(1) = 1,\ f(x) \leq f(y)$ for $x \leq y$, $(x,y \in S)\}$ would be topologically spanned by the convex combinations of its extreme points (Krein-Milman theorem). But the only extreme point in P is the constant function 1.

2.2: Let A be an infinite set, and let $S = A \cup \{0\}$ and have the one-point compactification topology (0 is the only limit point). Give S the semigroup structure of $s^2 = s$, $st = 0$ $(s,t \in S,\ s \neq t)$. Thus S is a semilattice, which, interestingly, has no discontinuous semicharacters. The algebra $\mathcal{R}(S) = \ell^1(A) \oplus \mathbb{C}$:

Let $p \in A$, then χ_p defined by $\chi_p(s) = 0$, $s \neq p$, $\chi_p(p) = 1$, is a continuous semicharacter of S, and the sums of the form $c_0 + \Sigma_{p \in A} c_p \chi_p$ with $\Sigma_p |c_p| < \infty$ are exactly $\ell^1(A) \oplus \mathbb{C}$. Thus Proposition 1.7 shows that $\ell^1(A) \oplus \mathbb{C} \subset \mathcal{R}(S)$.

Let $f \in \mathcal{R}(S)$ and so there exists $(T,\mu,\Omega) \in \mathcal{S}$ with $f(x) = \int_\Omega (Tx)\, g\, d\mu$, $x \in S$.

Since S is an idempotent semigroup $Ts = \chi_{E_s}$, $s \in S$. By passing to the maximal ideal space Ω_1 of $L^\infty(\mu,\Omega)$ we may identify the measurable sets of positive measure with the nonempty open-closed subsets of Ω_1 so E_s is an open-closed subset of Ω_1. Further $E_0 \subset E_s$ for all $s \in S$, and $E_s \cap E_t = E_0$ for $s \neq t$. The sets $E_s \setminus E_0$ $(s \in A)$ are pairwise disjoint and have positive measure if nonempty. Thus at most countably many $E_s \neq E_0$.

2.2.3

We have

$$|f(s) - f(0)| \leq \int_{E_s \setminus E_0} |g| d\mu$$

and so

$$\Sigma_{s \in A} |f(s)-f(0)| + |f(0)| \leq ||g||_1$$

(only countably many nonzero terms in the sum). Thus $f \in \ell^1(A) \oplus \mathbb{C}$. The norm on $R(S)$ is given by

$$||f||_R = |f(0)| + \Sigma_{s \in A} |f(s)-f(0)|, \quad f \in R(S).$$

This example shows that a finitely supported function for a discrete commutative semigroup S need not in general be in $R(S)$; for example let $f = \chi_{\{0\}}$ on the above semigroup made discrete.

2.3 Remark: In Chapter 5, we will show for a locally compact abelian group G with dual group $\hat{G}$ that $R(G) = M(\hat{G})^{\wedge}$.

2.4: There exist semigroups S which have faithful representations with values in a W*-algebra but have no faithful L^∞-representations: let $S = [0,1]$ with $xy = \min(1,x+y)$. On $L^2([0,1])$, define $T_t f(x) = f(x+t)$ for $x+t < 1$, and $T_t f(x) = 0$ for $x+t \geq 1$ $(x,t \in [0,1],\ f \in L^2([0,1]))$. The operator $T_0 = 1$, $T_1 = 0$, and each T_t is nilpotent $(0 < t < 1)$ with $||T_t|| = 1$.

2.5 Remark: Most of the above results in Chapter 2 can be found in Dunkl and Ramirez [2].

2.3.1

§3. Examples of monothetic semigroups

A (necessarily commutative) topological semigroup S is called monothetic if it contains an element a (the generator) such that the set $\{a, a^2, \cdots\}$ is dense in S. For compact topological semigroups the set $G = \bigcap_{k=1}^{\infty} c\ell\{a^k, a^{k+1}, \cdots\}$ $(a \in S)$ is a closed abelian subgroup (see, for example, Hewitt and Ross [1,p. 100]). Thus a compact monothetic topological semigroup S is the union of the set $\{a, a^2, a^3, \cdots\}$ (a the generator of S) and a subgroup G. In particular, this means that S will have only one idempotent. We give an example to show that a (commutative) compact semitopological monothetic semigroup can have infinitely many idempotents.

3.1 Proposition: *If U is a unitary operator on a Hilbert space $\mathcal{H}$ with continuous spectrum, then the zero operator 0 is in the weak-operator closure of $\{U^n : n \in Z_+\}$.*

Proof. Since U is unitary we may write $U = \int e^{it} dE(t)$ (E the spectral measure of U, see Appendix B). By assumption, for $h_1, h_2 \in \mathcal{H}$, the measure $dm(t) = d\langle E(t)h_1, h_2\rangle$ is continuous. We claim that 0 is a cluster point of $\{\langle U^n h_1, h_2\rangle : n \in Z_+\}$. For if not,

$$|\langle U^n h_1, h_2\rangle| \geq C > 0 \quad \text{for } n \geq N, \text{ and}$$

$$|\langle U^{n+N} h_1, h_2\rangle| = |\langle U^n U^N h_1, h_2\rangle| \geq C > 0 \quad (n \in Z_+).$$

But the means of the measure dm(t) associated with $U^N h_1, h_2$ would satisfy

2.3.2

$$\frac{1}{n+1}\left(|\hat{m}(0)|^2+\cdots+|\hat{m}(n)|^2\right)=\frac{1}{n+1}\left(|\langle U^N h_1,h_2\rangle|^2+\cdots+|\langle U^{N+n}h_1,h_2\rangle|^2\right)$$

$$\geq c^2 > 0.$$

This implies by a theorem of Wiener (see Rudin [1,p. 117]) that dm(t) must have an atom, a contradiction. □

3.2 Remark: The semigroup of operators $\{U^n : n \in Z_+\}$ has been studied extensively by T. West [1], and the statement of Proposition 3.1 is contained in his Theorem 2.2.

3.3 Notation: Let the measure μ be a positive nonzero continuous measure on the circle group $\mathbb{T}$. Let $U : L^2(\mu) \to L^2(\mu)$ be the unitary operator defined by multiplication by z (that is, $Uf = zf(z)$, $f \in L^2(\mu)$, $z \in \mathbb{T}$). The weak-operator closure of $\{U^n : n \in Z_+\}$ in $\mathcal{B}(L^2(\mu))$ is denoted by $S_{wo}(\mu)$. The previous result shows that $S_{wo}(\mu)$ is a compact monothetic commutative semitopological semigroup possessing distinct idempotents 0 and 1.

3.4 Theorem: *Let μ be a continuous measure on the circle group whose support E is a Kronecker Cantor set, and let U be the multiplication operator by z on $L^2(\mu)$. The semigroup $S_{wo}(\mu)$ contains uncountably many idempotents.*

Proof. The existence of such a Kronecker Cantor set E may be found in Rudin [1, p. 103]. The set E has the property that any continuous unimodular function fon E can be uniformly approximated on E by a continuous character; that is, there exists $n \in Z$ with $|f(z)-z^n| < \varepsilon$ for $z \in E$ ($\varepsilon > 0$ given). We need

2.3.4

to show that the integer n may be chosen in Z_+. Let $\{f_k\}_{k=1}^{\infty}$ be a sequence of continuous unimodular functions on E with $f_k \neq 1$ and $\frac{1}{2k} < ||f_k - 1||_E < 1/k$. Choose $n_k \in Z$ with $||f_k - z^{n_k}||_E < 1/2k$. Thus $z^{n_k} \overset{k}{\rightarrow} 1$ and $n_k \neq 0$ (all k). By extracting a subsequence we assume either all $n_k > 0$ or all $n_k < 0$. Now note that in an abelian topological group with an identity (in particular, $c\ell\, Z$ the closure of $\{z^n : n \in Z\}$ in the sup-norm over E) if a net $\{x_\alpha\}$ converges to the identity 1, then $x_\alpha^{-1} \overset{\alpha}{\rightarrow} 1$ also. Thus we assume $n_k > 0$. Thus given f a continuous unimodular function on E let $\{n_\ell\}_{\ell=1}^{\infty}$ be a sequence from Z with $z^{n_\ell} \overset{\ell}{\rightarrow} f$. We assume either all $n_\ell < 0$ or $n_\ell > 0$. In the case that all $n_\ell < 0$, for each n_ℓ we find $n_{k,\ell}$ with $k \geq \ell$ and $n_\ell + n_{k,\ell} > 0$. But $z^{n_\ell + n_{k,\ell}} = z^{n_\ell} z^{n_{k,\ell}} \overset{\ell}{\rightarrow} f$, so we may assume the chosen sequence is from Z_+.

For $\xi \in E$ define f_ξ by restricting to the set E the function g given by

$$g(z) = \begin{cases} \bar{\xi} z, & 0 < \arg z \leq \arg \xi \\ 1, & \text{otherwise.} \end{cases}$$

Since a Kronecker set is independent (Rudin [1,p. 99]) $z = 1 \notin E$, and so f_ξ is a continuous unimodular function. Let $\{n_k\}_{k=1}^{\infty}$ be a sequence from Z_+ with $z^{n_k} \overset{k}{\rightarrow} f_\xi$ (uniformly on E). Hence for $h_1, h_2 \in L^2(\mu)$, $\int_E (z^{n_k} - f_\xi(z)) h_1(z) \bar{h}_2(z)\, d\mu(z) \overset{k}{\rightarrow} 0$, and so the operator F_ξ on $L^2(\mu)$ given by multiplication by f_ξ is in $S_{wo}(\mu)$.

2.3.5

Write $\mu = \mu_1 \oplus \mu_2$ where μ_1 is μ restricted to the set $E_1 = \{z \in E : f_\xi(z) = 1\}$ and μ_2 is μ restricted to the set $E_2 = \{z \in E : f_\xi(z) = \bar{\xi}z\}$. (Recall μ is a continuous measure.) Thus we write $L^2(\mu) = L^2(\mu_1) \oplus L^2(\mu_2)$, $U = U_1 \oplus U_2$, and $F_\xi = F_1 \oplus F_2$. From Theorem 3.1, there exists a sequence $\{n_\ell\} \subset Z_+$ with $U_2^{n_\ell} \overset{\ell}{\to} 0$ in $S_{wo}(\mu_2)$. For $z \in E_2$, $\xi f_\xi(z) = z$, and so $\xi F_2 = U_2$. Thus $F_2^{n_\ell} = \bar{\xi}^{n_\ell} U_2^{n_\ell} \overset{\ell}{\to} 0$ in $S_{wo}(\mu_2)$ (since $|\xi| = 1$).

For $z \in E_1$, $f_\xi(z) = 1$ and so $F_1 = 1$ in $S_{wo}(\mu_1)$. Thus $F_\xi^{n_\ell} = F_1^{n_\ell} \oplus F_2^{n_\ell} \overset{\ell}{\to} 0 \oplus 1$ in $S_{wo}(\mu)$; that is, the idempotent operator determined by multiplication by the characteristic function of E_1 is in $S_{wo}(\mu)$. □

3.5 Remark: The above theorem is due to Brown and Moran [1].

Chapter 3. Positive-Definite and Completely Positive Functions

Two of the most basic theorems in harmonic analysis on locally compact abelian groups are the Bochner theorem (Rudin [1, p. 19]) and the Eberlein theorem (Rudin [1, p. 32]). These theorems characterize the positive-definite functions, and the space of Fourier-Stieltjes transforms, respectively. In this chapter we extend these results to semigroups with a dense union of groups (called semigroups of type $\mathcal{U}$) in Theorems 1.5 and 2.6. Given a positive-definite function it is possible to construct a Hilbert space representation which is equivalent to an L^∞-representation. Our methods lead us to study completely monotone functions in Theorem 3.13 - an extension of the Hausdorff moment problem. The underlying idea is that of a positive semicharacter.

§1. Positive-definite functions on semigroups with a dense union of groups

The concept of a positive-definite function extends to commutative semigroups which have a dense union of groups (semigroups of type $\mathcal{U}$). The analogue of the classical theorem of Bochner showing the equivalence of positive-definite functions on a locally compact abelian group G and Fourier-Stieltjes transforms of positive measures on the dual group $\hat{G}$ of G is given for commutative semigroups of type $\mathcal{U}$.

1.1 Definition: Let S be a commutative semigroup which has a dense subsemigroup U which is a union of groups. The semigroup S is said to be of type $\mathcal{U}$. For each $x \in U$, there exists a unique element $x' \in U$ such that $e = xx'$ is an idempotent and $ex = x$, $ex' = x'$. If $(T,\mu,\Omega) \in \mathcal{S}$, then $(Tx)(Tx')$ is an idempotent in $L^\infty(\mu,\Omega)$ and thus $|Tx| = 0$ or 1 (μ-a.e.) and $Tx' = \overline{Tx}$.

1.2 Definition: Let S be of type $\mathcal{U}$ and suppose S possesses an identity 1. For $f \in C^B(S)$, f is said to be positive-definite if and only if for every finite subset $x_1,\cdots,x_N \in U$ and $c_1,\cdots,c_N \in \mathbb{C}$, we have

$$(P_1) \qquad \Sigma_{i,j=1}^{N} c_i \bar{c}_j \; f(x_i \; x_j') \geq 0.$$

1.3 Definition: The space $R_+(S)$ is the subspace of $R(S)$ consisting of those f for which $f = T^*g$, $g \geq 0$, $g \in L^1(\mu)$, $(T,\mu,\Omega) \in \mathcal{S}$.

1.4 Proposition: Let S be a semigroup of type $\mathcal{U}$ with an identity, and let $f \in R_+(S)$. Then f is positive-definite.

Proof. Let $x_1,\cdots,x_N \in U$ and $c_1,\cdots,c_N \in \mathbb{C}$. Then

$$\Sigma_{i,j=1}^{N} c_i\bar{c}_j f(x_i x_j') = \int_\Omega \Sigma_{i,j=1}^{N} c_i\bar{c}_j (Tx_i Tx_j') g d\mu$$

$$= \int_\Omega |\Sigma_{i=1}^{N} c_i (Tx_i)|^2 \, gd\mu \geq 0$$

since $g \geq 0$. □

1.5 Theorem: Let S be a semigroup of type $\mathcal{U}$ with an identity 1. If $f \in C^B(S)$ is positive-definite, then $f \in R_+(S)$.

Proof. The proof is similar to the analogous topological group result due to Gelfand and Raikov (Naimark [1, p. 393]).

3.1.5

Let f be positive-definite on S. For e an idempotent in S, $0 \leq f(ee') = f(ee) = f(e)$. In particular, $f(1) \geq 0$. For $x \in U$ with $xx' = e$ and $\lambda \in \mathbb{C}$,

$$(*) \quad \begin{aligned} f(ee') + \bar{\lambda}f(ex') + \lambda f(e'x) + |\lambda|^2 f(xx') &\geq 0, \\ f(e) + \bar{\lambda}f(x') + \lambda f(x) + |\lambda|^2 f(e) &\geq 0; \end{aligned}$$

thus $\bar{\lambda}f(x') + \lambda f(x)$ is real for all $\lambda \in \mathbb{C}$.

In particular, $f(x') + f(x)$ and $\bar{i}f(x') + if(x) = i(-f(x') + f(x))$ are real. This implies $f(x') = \overline{f(x)}$. For any $x \in U$, we claim $|f(x)| \leq f(1)$: if $f(1) = 0$, then in (*) with $e = 1$ and $\lambda = -\overline{f(x)}$, we have

$$\begin{aligned} -f(x)f(x') - \overline{f(x)}f(x) &\geq 0, \\ -|f(x)|^2 - |f(x)|^2 &\geq 0, \end{aligned}$$

and so $|f(x)| = 0$; if $f(1) \neq 0$, then let $\lambda = -\overline{f(x)}/f(1)$, to yield

$$f(1) - \frac{f(x)f(x')}{f(1)} - \frac{\overline{f(x)}f(x)}{f(1)} + \frac{|f(x)|^2 f(1)}{f(1)^2} \geq 0,$$

$$f(1) - \frac{|f(x)|^2}{f(1)} - \frac{|f(x)|^2}{f(1)} + \frac{|f(x)|^2}{f(1)} \geq 0,$$

$$f(1) \geq \frac{|f(x)|^2}{f(1)}, \quad f(1) \geq |f(x)|.$$

We assume $f \neq 0$. Let A be the $*$-algebra $\{\xi = \Sigma_{i=1}^N c_i\delta_{x_i} : c_i \in \mathbb{C},\ x_i \in U\}$ where $\delta_x\delta_y = \delta_{xy}$ and $(\Sigma_{i=1}^N c_i\delta_{x_i})^* = \Sigma_{i=1}^N \bar{c}_i\delta_{x_i'}$ (δ_x is the unit point mass at $x \in S$). On A define the bilinear form $\langle\Sigma_{i=1}^N c_i\delta_{x_i}, \Sigma_{j=1}^M d_j\delta_{y_j}\rangle = \Sigma_{i,j} c_i\bar{d}_j f(x_iy_j')$. Note the Cauchy-Schwartz inequality

3.1.5

$|<\xi,\eta>|^2 \leq <\xi,\xi><\eta,\eta>$ $(\xi,\eta \in A)$. Let N be the ideal in A of those $\xi \in A$ with $<\xi,\xi> = 0$. Then A/N is an inner product space and we denote its Hilbert space completion by H.

For each $x \in U$, define the operator Tx on A by $Tx(\Sigma\, c_i\delta_{x_i}) = \Sigma c_i\delta_{xx_i}$ $(\Sigma\, c_i\delta_{x_i} \in A)$. The problem is to show that Tx is a contraction operator. For this we begin by letting $x \in U$, $\xi = \Sigma c_i\delta_{x_i} \in A$, and $\eta = \Sigma d_j\delta_{y_j} \in A$. Then $<Tx(\xi),Tx(\eta)> = \Sigma\, c_i\bar{d}_j\, f(xx_i(xy_j)') = \Sigma c_i\bar{d}_j\, f(ex_iy_j') = <Te(\xi),\eta>$ where $e = xx'$, an idempotent in U.

Consider the map $Te:A \to A$ defined by $Te(\Sigma\, c_i\delta_{x_i}) = \Sigma c_i\delta_{ex_i}$. Note that $Te = Te^*$ and $Te^2 = Te$. Now applying the Cauchy-Schwartz inequality, we have for $\xi,\eta \in A$ that

$$|<Te\xi,Te\eta>|^2 = |<Te\xi,\eta>|^2$$

$$\leq <Te\xi,Te\xi><\eta,\eta> \quad ;$$

in particular, $|<Te\xi,Te\xi>| \leq <\xi,\xi>$. It follows that $||Te|| \leq 1$.

Hence for $x \in U$, $||Tx|| \leq 1$ and so it extends uniquely to H: indeed for $\xi,\eta \in A$,

$$<Tx(\xi),Tx(\eta)> = |<Txx'\xi,\eta>|$$

$$\leq ||Txx'||\ |<\xi,\eta>| \leq |\xi,\eta>|.$$

Note that $T(xy) = (Tx)(Ty)$, $(x,y \in U)$.

For $x \in U$, $(Tx)^* = Tx'$: consider

3.1.5

$$\langle Tx\, \Sigma c_i \delta_{x_i}, \Sigma\, d_j \delta_{y_j} \rangle = \Sigma_{i,j}\; c_i \overline{d}_j\; f(xx_i y_j')$$

$$= \langle \Sigma\, c_i \delta_{x_i}, T(x')\Sigma\, d_j \delta_{y_j} \rangle, \quad (x_i, y_j \in U).$$

Thus $\{Tx : x \in U\}$ is a commutative set of normal contraction operators on H .

Let $x \in S$, then there exists a net $\{x_\alpha\} \subset U$ such that $x_\alpha \to x$. Since the unit ball in $B(H)$ is compact in the weak operator (WO) topology (Dunford and Schwartz [1, p. 511]), there exists a cluster point T_∞ of $\{Tx_\alpha\}$. But for $\Sigma\, c_i \delta_{x_i}, \Sigma\, d_j \delta_{y_j} \in A$,

$$\langle T_\infty\, \Sigma\, c_i \delta_{x_i}, \Sigma\, d_j \delta_{y_j} \rangle$$

$$= \lim_\alpha \langle Tx_\alpha\, \Sigma\, c_i \delta_{x_i}, \Sigma\, d_j \delta_{y_j} \rangle$$

$$= \lim_\alpha \Sigma_{i,j}\; c_i \overline{d}_j\; f(x_\alpha x_i y_j')$$

$$= \Sigma_{i,j}\; c_i \overline{d}_j\; f(xx_i y_j') \quad .$$

For $x \in S$, we define $Tx = T_\infty$. From the above we see that this is well-defined and that $x \mapsto Tx$ is a continuous map of S into $(B(H), WO)$ since $||Tx|| \leq 1$ and A is dense in H . For $x \in S$ and $u \in U$, the separate continuity of the WO-topology shows that $Txu = TxTu$: since for $x_\alpha \to x(x_\alpha \in U)$,

3.1.5

$$\langle Txu \textstyle\sum c_i\delta_{x_i}, \sum d_j\delta_{y_j}\rangle \quad (x_i, y_j \in U)$$

$$= \lim_\alpha \langle Tx_\alpha u \textstyle\sum c_i\delta_{x_i}, \sum d_j\delta_{y_j}\rangle$$

$$= \lim_\alpha \textstyle\sum c_i\bar{d}_j\ f(x_\alpha u x_i y_j')$$

$$= \textstyle\sum c_i\bar{d}_j\ f(xux_i y_j')$$

$$= \langle Tx \textstyle\sum c_i\delta_{ux_i}, \sum d_j\delta_{y_j}\rangle$$

$$= \langle TxTu \textstyle\sum c_i\delta_{x_i}, \sum d_j\delta_{y_j}\rangle \quad .$$

Similarly, for $x,y \in S$, $Txy = TxTy$; and so $x \mapsto Tx$ is a continuous representation of S into the unit ball of $B(H)$. As $x_\alpha \to x$ $(x_\alpha \in U)$, $Tx_\alpha^* \to Tx^*$ so $Tx^*Tx = \text{WO-lim}_\alpha Tx_\alpha{}^*Tx = \text{WO-lim}_\alpha TxTx_\alpha{}^* = TxTx^*$. Thus $\{Tx:x \in S\}$ is a commutative semigroup of normal contraction operators.

We next observe that $\delta_1 \in A$ is a cyclic vector for $\{Tx:x \in S\}$ and that

$$f(x) = \langle (Tx)\delta_1, \delta_1\rangle, \ x \in S.$$

The final step is to apply the spectral theorem for the given cyclic representation (Appendix B) of the norm-closed algebra in $B(H)$ generated by $\{Tx:x \in S\}$. We thus obtain a probability measure μ on the maximal ideal (compact Hausdorff) space Ω of the algebra, and an isometry j of H onto $L^2(\mu)$ such that for $x \in S$, $\xi,\eta \in H$, $j(Tx\xi) = (\hat{Tx})j\xi$ and $\langle (Tx)\xi,\eta\rangle = \int_\Omega (\hat{Tx})j\xi\overline{j\eta}d\mu$. Here $\hat{Tx}$ is the Gelfand transform of Tx, so that $x \mapsto \hat{Tx}$ is an L^∞-representation of S into $L^\infty(\mu)$.

Observe that the weak-* topology on $L^\infty(\mu)$ is the same as the WO-topology as operators on $L^2(\mu)$. Also

$$f(x) = \langle (Tx)\delta_1, \delta_1 \rangle = \int_\Omega (\hat{Tx}) \, |j\delta_1|^2 d\mu,$$

thus $f \in R_+(S)$. □

1.6 Remark: Let $S = Z_+ \cup \{\infty\}$ where $nm = \infty$ for $n \neq m$ and $nn = n$. Then $R(S) = \ell^1(Z_+) \oplus \mathbb{C}$ (2.2.2). Note that S is of type U but M(S) has no identity. A function $f \in C^B(S)$ will be in $R_+(S)$ if and only if $f(n) \geq f(\infty) \geq 0$, $(n \in Z_+)$ and $\Sigma_{n=1}^{\infty}(f(n)-f(\infty)) < \infty$. A short argument shows that $f \in C^B(S)$ satisfies (P_1) if and only if $f(n) \geq f(\infty) \geq 0$, $(n \in Z_+)$, and so the requirement that S has an identity in Theorem 1.5 is crucial: Let A be an $N \times N$ matrix such that $a_{ij} = a_i$ for $i = j$ and $a_{ij} = a_1$ for $i \neq j$. Thus

$$A = \begin{bmatrix} a_1 & a_1 & a_1 & \cdots & a_1 \\ a_1 & a_2 & a_1 & \cdots & a_1 \\ a_1 & a_1 & a_3 & \cdots & a_1 \\ \cdot & \cdot & \cdot & \cdot & \cdot \\ a_1 & a_1 & \cdots & & a_N \end{bmatrix} .$$

It is easy to check the determinants of the upper left truncates of A are nonnegative if and only if $a_n \geq a_1 \geq 0$ $(1 \leq n \leq N)$. (Recall that a Hermitian matrix is positive-definite if and only if the principal minors are all positive.)

1.7 Remark: If S does not possess an identity, then in Definition 1.2, we require f to satisfy the following additional condition:

$$(P_2) \qquad |\Sigma_{i=1}^{N} c_i\ f(x_i)\ |^2 \leq K\Sigma_{i,j}\ c_i\bar{c}_j\ f(x_i x_j'), (x_i, x_j \in U).$$

Note f already satisfies $f(x') = \overline{f(x)}$ for $x \in U$.

For then one can adjoin an identity 1 to S to obtain S_1 and set $f(1) = K$. The function f is now positive-definite on S_1: since for $1 = x_0, x_1, \cdots, x_N \in U$ and $c_0, c_1, \cdots, c_N \in \mathbb{C}$,

$$\Sigma_{i,j=0}^{N} c_i\bar{c}_j\ f(x_i x_j')$$

$$= |c_0|^2 K + 2\mathrm{Re}(c_0 \overline{\Sigma_{i=1}^{N}\ c_i\ f(x_i)}) + \Sigma_{i,j=1}^{N} c_i\bar{c}_j\ f(x_i x_j')$$

$$\geq |c_0|^2 K - 2|c_0|\,|\Sigma_{i=1}^{N}\ c_i\ f(x_i)| + \Sigma_{i,j=1}^{N} c_i\bar{c}_j\ f(x_i x_j')$$

$$\geq (|c_0|K^{1/2} - (\Sigma_{i,j=1}^{N} c_i\bar{c}_j\ f(x_i x_j'))^{1/2})^2 \geq 0.$$

This condition is the usual condition for a positive functional on a $*$-algebra to be extendable to a positive functional on the $*$-algebra with an identity adjoined, (Hewitt and Ross [1, p. 317].)

We describe two important examples of semigroups of type $\mathcal{U}$.

1.8 <u>Lemma</u>: <u>Let</u> μ <u>be a continuous probability measure on the measure space</u> Ω <u>and let</u> t <u>be a real number with</u> $0 < t < 1$. <u>There exists a measurable subset</u> $F \subset \Omega$ <u>with</u> $\mu(F) = t$.

Proof. Let ε be such that $0 < \varepsilon < t$. Write $L^\infty(\mu,\Omega) \cong C(X)$, X a compact Stonean space, (see Gamelin [1, p. 17].) The open-closed subsets of X correspond to the nonzero measurable subsets of Ω (modulo μ-null sets). For each $p \in X$, let V_p be an open-closed neighborhood of p with $\mu(V_p) < \varepsilon$. By the compactness of X, there exists a finite subcover $\{V_{p_1},\cdots,V_{p_n}\}$ of X. Choose a partial union E_1 of the finite subcover so that $t-\varepsilon < \mu(E_1) \leq t$. Similarly, choose a subset $E_2 \subset X \setminus E_1$ with $t-\varepsilon/2 < \mu(E_1) + \mu(E_2) \leq t$. Continue by induction to find subsets $E_{n+1} \subset X \setminus (E_1 \cup \cdots \cup E_n)$ with $t-\varepsilon/2^n < \mu(E_1) +\cdots+\mu(E_{n+1}) \leq t$. Let $E = \bigcup_{n=1}^\infty E_n$. There exists an open-closed subset E' of X with $\chi_{E'} = \chi_E$ μ- a.e., and E' corresponds to a desired subset $F \subset \Omega$. □

For a more general result see Rudin [2, p. 114].

1.9 <u>Example</u>: Given a continuous probability measure μ on the measure space Ω, the semigroup S_μ is defined to be the unit ball of $L^\infty(\mu,\Omega)$ under multiplication and with the weak-* topology. Let $U = \{f \in S_\mu : |f| = 1 \text{ or } 0\ \mu\text{-a.e.}\}$. Thus U is the union of all the subgroups of S_μ. We assert that U is dense in S_μ and so S_μ is a compact semitopological semigroup of type $\mathcal{U}$:

(1) Fix a positive $\lambda < 1$. Let $P = \{E_1,\cdots,E_n\}$ be a finite partition of Ω of disjoint μ-measurable subsets. For each E_i $(1 \leq i \leq n)$ let $E_{i,\lambda} \subset E_i$ with $\mu(E_{i,\lambda}) = \lambda\mu(E_i)$ (by Lemma 1.8). Define χ_P to be the characteristic function of the set $\bigcup_{i=1}^n E_{i,\lambda}$. The net $\{\chi_P : P \text{ a finite partition}\}$ has as a weak-* cluster point in S_μ the constant function λ. Thus $\lambda \in \overline{U}$.

(2) Let $f = \lambda\chi_E$ (E a measurable subset of Ω, $|\lambda| \leq 1, \lambda \in \mathbb{C}$).

There exists a net $\{g_\alpha\} \subset U$ with $g_\alpha \overset{\alpha}{\to} |\lambda|$. Thus $f = \lim_\alpha g_\alpha(\operatorname{sgn} \lambda)\chi_E \in \bar{U}$.

(3) Finally, any $h \in S_\mu$ is a limit of finite sums of multiples of characteristic functions with disjoint supports. □

1.10 Remark: Example 1.9 furnishes us with an example of a compact semitopological semigroup of type $\mathcal{U}$ for which the map $x \mapsto x^2$ is not continuous, indeed converge to the constant function 1/2 with characteristic functions.

1.11 Example: Another useful example of semigroups of type $\mathcal{U}$ is given by piecing locally compact abelian (LCA) groups together. Let $\{G_\alpha\}_{\alpha \in \Lambda}$ be a finite collection of LCA groups with continuous homomorphisms $\pi_{\alpha\beta}: G_\alpha \to G_\beta$ $(\alpha, \beta \in \Lambda, \alpha > \beta)$ where Λ is a finite (lower) semilattice, and $\alpha > \beta > \gamma$ implies $\pi_{\beta\gamma} \circ \pi_{\alpha\beta} = \pi_{\alpha\gamma}$ $(\alpha, \beta, \gamma \in \Lambda)$. Define $S = \bigcup_{\alpha \in \Lambda} G_\alpha$. The multiplication in S is given by (1) for $x, y \in G_\alpha$, $xy = x+y \in G_\alpha (\alpha \in \Lambda)$, and (2) for $x \in G_\alpha$, $y \in G_\beta$, $\alpha \neq \beta$, let $\gamma = \alpha \wedge \beta$ and $xy = \pi_{\alpha\gamma} x + \pi_{\beta\gamma} y \in G_\gamma$. To topologize S, for $\alpha \neq 0$ (0 the minimum element in Λ) use the topology from G_α. For $x \in G_0$, choose a relatively compact neighborhood W of x in G_0 and in each G_α $(\alpha > 0)$ pick a compact set K_α, then $V = W \cup \bigcup\{\pi_{\alpha 0}^{-1} W \setminus K_\alpha : \alpha > 0\}$ is a basic neighborhood of $x \in S$. If G_0 is compact, then S is a compact semitopological semigroup. We compute $\mathcal{R}(S)$ in Example 5.1.7.

Let $G_1 = H$ be any LCA group and $G_0 = \{0\}$, $\pi_{10}: G_1 \to G_0$, and $\Lambda = \{0,1\}$. Then $S = \bigcup_{\alpha \in \Lambda} G_\alpha$ is the one-point compactification of H. For the infinite version of this construction, see J. Berglund, Semigroup Forum 5(1973), 191-215.

3.1.12

We end this section by showing that commutative compact semitopological semigroups of type $\mathcal{U}$ give the universal object in which semigroups whose representation algebra separates the points of the semigroup can be faithfully represented.

1.12 Theorem: *Let S be a commutative semitopological semigroup with representation algebra $\mathcal{R}(S)$ separating the points of S. Then there exists a commutative compact semitopological semigroup of type $\mathcal{U}$ which contains a faithful homomorphic image of S.*

Proof. For $(T,\mu,\Omega) \in \mathcal{S}$, decompose $\mu = \mu_c + \mu_d$ into its continuous and discrete pieces. Let S_1 be the unit ball of the direct sum $\Sigma \oplus L^\infty(\mu_c)$ (sum over all the continuous pieces, see 2.1.10), with the coordinate weak-* topology. By 1.9, S_1 is of type $\mathcal{U}$ and let $\pi_1 : S \to S_1$ by $x \mapsto (Tx)_{\mu_c}$, $(T,\mu,\Omega) \in \mathcal{S}$.

Let $(T,\mu,\Omega) \in \mathcal{S}$, $\mu = \mu_c + \mu_d$, and $\omega \in \text{spt } \mu_d$ (the support of μ_d). Define $\chi^{(\mu,\omega)}(x) = \chi(x) = (Tx)(\omega)$, $(x \in S)$. The function $x \mapsto \chi(x)$ is a continuous semicharacter of S. If $|\chi(x)| = 1$ or 0 for all $x \in S$, we call χ a T_0-semicharacter. The semigroup $\bar{\mathbb{T}} \cup \{0\}$ is of type $\mathcal{U}$. Let $S_2 = \Pi(\bar{\mathbb{T}} \cup \{0\})$ (product over all T_0-semicharacters of S) and define $\pi_2 : S \to S_2$ by $x \mapsto (\chi(x))_\chi$, χ a T_0-semicharacter of S.

Let $(T,\mu,\Omega) \in \mathcal{S}$, $\mu = \mu_c + \mu_d$, $\omega \in \text{spt } \mu_d$, and suppose $0 < |\chi^{(\mu,\omega)}(y)| = |\chi(y)| < 1$ for some $y \in S$. Let λ denote the normalized arc length measure on the circle in $\mathbb{C}$ with center 1 and radius 1/2, so $\int_{\mathbb{C}} f\, d\lambda = \frac{1}{2\pi}\int_0^{2\pi} f(1+e^{i\theta}/2)\, d\theta$, $f \in C_0(\mathbb{C})$. The function $f(z,x) = \operatorname{sgn} \chi(x)|\chi(x)|^z$ defined on $\{z \in \mathbb{C}: Re\ z > 0\} \times S$ is analytic in z, and the map $\pi_3: S \to L^\infty(\lambda)$ defined by $\pi_3(x) = f(\cdot,x)$ maps S weak-* continuously into the unit ball S_λ of $L^\infty(\lambda)$, $(x \in S)$. Furthermore, $\chi(x) = \int_C f(z,x)\, d\lambda(z)$ by the Poisson integral formula. Let $S_3 = \Pi S_\lambda$ (product over all such χ,μ,ω) and let $\pi_3: S \to S_3$ by $\pi_3 x = (f(\cdot,x))$.

The map $\pi: S \to S_1 \times S_2 \times S_3$ defined by $\pi = \pi_1 \otimes \pi_2 \otimes \pi_3$ is a continuous homomorphism, and π is faithful since given any $f \in R(S)$ there exist functions $f_i \in R(S_i)$ $(i = 1,2,3)$ with $f = (f_1\circ\pi_1) + (f_2\circ\pi_2) + (f_3\circ\pi_3)$ and by hypothesis $R(S)$ separates the points of S. □

§2. The dual of $R(S)$

2.1 Notation: The dual space of $R(S)$ is denoted by $A(S)$.

2.2 Remark: For semigroups S of type U, $A(S)$ is a commutative W*-algebra (see 1.1.15).

2.3 Definition: For $x \in S$, define $\rho x \in A(S)$ by $\rho x(f) = f(x)$, $f \in R(S)$.

2.4 Theorem: Let S be a semigroup of type U, and let T be a representation of U into the unit ball of $C(\Omega)$, Ω compact Hausdorff (no continuity assumed for T). Suppose there exists $\mu \in M(\Omega)$ such that the function $\phi: x \to \int_\Omega (Tx)\, d\mu$ on U is the restriction of a function in $C^B(S)$. Then there exist an L^∞-representation $(T_1,\mu_1,\Omega_1) \in S$ and $g \in L^1(\mu_1)$ such that $\phi(x) = T_1^* g(x)$, $x \in U$; that is, ϕ is a restriction to U of a function $\psi \in R(S)$. Furthermore $||\psi||_R \leq ||\mu||$.

Proof. We form the compact space Ω_1 by identifying the points of the support of μ under the equivalence $w_1 \sim w_2$ if and only if $Tx(w_1) = Tx(w_2)$ for all $x \in U$. Now define $T_1 x \in C(\Omega_1)$ where $T_1 x$ agrees with Tx on these equivalence classes. Consider the canonical map $i:C(\Omega_1) \to C(\Omega)$, and its adjoint $i^*:M(\Omega) \to M(\Omega_1)$. Let $\mu' = i^*\mu$ and write $\mu_1 = |\mu'|$; let $g = \frac{d\mu'}{d\mu_1} \in L^1(\mu_1)$. Then $||g||_1 \leq ||\mu||$. (We may assume $||\mu_1|| = 1$.)

For $h \in L^1(\mu_1)$, define $\alpha_h(x) = \int_{\Omega_1} (T_1x) h d\mu'$, $x \in U$. For $h = \Sigma_{i=1}^N c_i T_1 x_i$ $(x_i \in U)$, α_h is a restriction to U of a function from $C^B(S)$. We now show this holds for all $h \in L^1(\mu_1)$.

For $h \in L^1(\mu_1)$, let $h_n \to h$ where h_n has the form $\Sigma\, c_i T_1 x_i$ (by the Stone-Weierstrass theorem the set $\{\Sigma\, c_i T_1 x_i : x_i \in U\}$ is dense in $C(\Omega_1)$, which is dense in $L^1(\mu_1)$). As $h_n \to h$ in $L^1(\mu_1)$, $\alpha_{h_n} \to \alpha_h$ uniformly on U (which is dense in S). Thus α_h is the restriction to U of a function from $C^B(S)$, $(h \in L^1(\mu_1))$. In particular, the map $x \mapsto T_1 x$ from U to $L^\infty(\mu_1)$ is a (weak-* continuous) L^∞-representation of U.

Let $x \in S$ and choose $\{x_\alpha\} \subset U$ with $x_\alpha \to x$. The set $\{T_1 x_\alpha\}$ has a weak-* cluster point which we denote by $T_1 x$. It is unique since for $h \in L^1(\mu_1)$, $y \mapsto \int_{\Omega_1} (T_1 y) h d\mu_1$ $(y \in U)$ is the restriction to U of a unique function in $C^B(S)$. This also shows that $x \mapsto T_1 x$

from S to $L^\infty(\mu_1)$ is weak-* continuous. Thus T_1 is a L^∞-representation of S, and $T_1^* g(x) = \phi(x)$ for $x \in U$. □

2.5 Corollary: Let S be of type $\mathcal{U}$ and let $\phi \in A(S)^*$. If $\phi : x \mapsto \phi(\rho x)$ is continuous on S, then $\phi \in R(S)$ and $||\phi||_R \leq ||\phi||$.

Proof. If S is of type $\mathcal{U}$, then $A(S) = C(\Omega)$, Ω hyper-Stonean. The map $x \mapsto \rho x$ is a representation of S into $C(\Omega)$, (2.1.11). Let $\mu \in M(\Omega)$ be such that $\phi(x) = \int_{\Omega_1} (\rho x) d\mu \quad (x \in S)$, $||\mu|| = ||\phi||$. By Theorem 2.4, $\phi \in R(S)$. Clearly $||\phi||_R \leq ||\mu|| = ||\phi||$. □

2.6 Theorem: Let S be of type $\mathcal{U}$. Then for $f \in C^B(S)$, $f \in R(S)$ with $||f||_R \leq K$ if and only if for $c_i \in \mathbb{C}$, $x_i \in S$,

(*) $$|\Sigma_{i=1}^N c_i f(x_i)| \leq K||\Sigma_{i=1}^N c_i \rho x_i||_A ,$$

(K a constant depending only on f).

Proof. Define $\phi(\Sigma_{i=1}^N c_i \rho x_i) = \Sigma_{i=1}^N c_i f(x_i)$. Then ϕ extends to $A(S)$ with $||\phi|| \leq K$ by the Hahn-Banach theorem. Since $x \mapsto \phi(\rho x) = f(x)$ is continuous, $f \in R(S)$ with $||f||_R \leq ||\phi|| \leq K$ by Corollary 2.5. □

2.7 Remark: By the Kaplansky density theorem (1.2.7), the unit ball of the algebra $\{\Sigma c_i \rho x_i : x_i \in U\}$ is weak-* dense in $A(S)$. Thus in condition (*) one only needs to consider $\Sigma c_i \rho x_i$ with $x_i \in U$. Also for $f \in R(S)$,

3.2.11

$$||f||_{R(S)} = \sup\{|\sum_{i=1}^{N} c_i f(x_i)| : ||\Sigma c_i \rho x_i|| \leq 1, x_i \in U, c_i \in \mathbb{C}\}$$

$$= ||f||_{R(U)} \quad .$$

2.8 Corollary: Let S be of type $\mathcal{U}$ and let $f \in C^B(S)$. Suppose $\{f_\alpha\} \subset R(S)$, $||f_\alpha||_R \leq K$, and $f_\alpha \to f$ pointwise on S. Then $f \in R(S)$ and $||f||_R \leq K$.

Proof. The function f satisfies condition (*) of Theorem 2.6. □

2.9 Remark: For locally compact abelian groups, Theorem 2.6 is the well-known Eberlein characterization of the Fourier-Stieltjes transforms (see Appendix A).

2.10 Definition: For S a semitopological semigroup, the semigroup S with the discrete topology is denoted by S_d.

2.11 Theorem: Let S_1 and S_2 be of type $\mathcal{U}$, and let $j: S_1 \to S_2$ be a one-to-one homomorphism with dense range. Then $R(S_1) \cap C^B(S_2) = R(S_2)$. For $f \in R(S_2)$, $||f||_{R(S_2)} = ||f|S_1||_{R(S_1)}$.

Proof. We may assume S_1 is a dense subsemigroup of S_2 with the topology on S_1 being stronger than that of S_2. The restriction $j^*: R(S_2) \to R(S_1)$ is one-to-one, and so $R(S_2) \subset C^B(S_2) \cap R(S_1)$. Clearly for $f \in R(S_2)$, $||f|S_1||_{R(S_1)} \leq ||f||_{R(S_2)}$.

3.2.12

Let $U_1 \subset S_1$ be a dense subsemigroup of S_1 which is a union of groups, let $f \in R(S_1) \cap C^B(S_2)$, and $\varepsilon > 0$. Thus there exists $(T,\mu,\Omega) \in S(S_1)$ with $T^*g = f$, $g \in L^1(\mu)$, and $||f||_R \geq ||g||_1 - \varepsilon$. Thus $x \mapsto Tx$ is a representation of U_1 into $L^\infty(\mu)$, and the function $\phi: x \mapsto \int_\Omega (Tx)\, g d\mu$ on U_1 agrees with f and so is a restriction to U_1 of a function in $C^B(S_2)$.

Theorem 2.4 now asserts that ϕ is a restriction to U_1 of a function in $R(S_2)$. But $\phi = f$ on U_1, and so $f \in R(S_2)$ and $||f||_{R(S_2)} \leq ||g||_1 \leq ||f||_{R(S_1)} + \varepsilon$. □

2.12 <u>Corollary</u>: <u>If</u> S <u>is of type</u> U, <u>then</u> $R(S_d) \cap C^B(S) = R(S)$.

Proof. Consider $U_d \to S$. By Theorem 2.11, $R(U_d) \cap C^B(S) = R(S)$. Thus $R(S_d) \cap C^B(S) \subset R(U_d) \cap C^B(S) = R(S)$. The other inclusion is clear. □

§3. Completely monotone functions

The Hausdorff one-dimensional moment problem (see Shohat and Tamarkin [1]) is the following: given a prescribed set of real numbers $\{v_n\}_{n=0}^\infty$, find a bounded nondecreasing function $\psi(x)$ on the closed interval [0,1] such that its moments are equal to the prescribed values; that is,

$$\int_0^1 t^n \, d\psi(t) = v_n, \quad (n \in Z_+).$$

The integral is a Riemann-Stieltjes integral. Equivalently, find a nonnegative measure μ on [0,1] with

$$\int_{[0,1]} t^n \, d\mu(t) = v_n, \quad (n \in Z_+).$$

3.3.2

3.1 Definition: Define the operator $\Delta^k (k = 0,1,2,\cdots)$ by

$$\Delta^0 v_n = v_n,$$

$$\Delta^1 v_n = v_n - v_{n+1},$$

$$\Delta^k v_n = v_n - \binom{k}{1} v_{n+1} + \binom{k}{2} v_{n+2} - \cdots - (-1)^k v_{n+k},$$

$(n \in Z_+)$ for any sequence of real numbers $\{v_n\}_{n=0}^{\infty}$. If $\Delta^k v_n \geq 0$ $(n \in Z_+)$, the sequence $\{v_n\}_{n=0}^{\infty}$ is called a classically completely monotone sequence.

We now state the extension (see 3.6) of this definition to commutative semitopological semigroups with identities.

3.2 Definition: On a commutative semitopological semigroup S with identity 1, for each $n \in Z_+$, define the operator Δ_n on the space $C^R(S)$ of continuous real-valued functions inductively by

$$\Delta_0 f(x) = f(x),$$

and

$$\Delta_n f(x;h_1,\cdots,h_n) = \Delta_{n-1} f(x;h_1,\cdots,h_{n-1})$$

$$- \Delta_{n-1} f(xh_n;h_1,\cdots,h_{n-1}),$$

$(f \in C^R(S),\ x,h_1,\cdots,h_n \in S,\ n = 1,2,\cdots)$. A function $f \in C^R(S)$ is said to be completely monotone if and only if $\Delta_n f \geq 0$ $(n \in Z_+)$. The space of such f is denoted by CM(S). Recall f_x denotes the translate of $f \in C(S)$ by $x \in S$; that is, $f_x(y) = f(xy)$, $y \in S$.

3.3.3

3.3 Definition: Let $\phi \in \hat{S}$ (S a commutative semitopological semigroup) with $\phi \neq 0$ and $0 \leq \phi(x) \leq 1$, $x \in S$. The semicharacter ϕ is said to be a positive semicharacter; and the space of such is denoted by $\hat{S}_+$.

3.4 Proposition: *Let* $\phi \in \hat{S}_+$, *then* $\phi \in CM(S)$.

Proof. Let $\phi \in \hat{S}_+$. By definition, $\Delta_0\phi(x) = \phi(x) \geq 0$ $(x \in S)$. Use induction to note that

$$\Delta_n\phi(x;h_1,\cdots,h_n) = \phi(x)\ \Pi_{i=1}^n(1-\phi(h_i)) \geq 0.$$

Thus $\phi \in CM(S)$. □

3.5 Lemma: *Let* X *be a set of real-valued functions on* Z_+ *with*

(1) $0 \leq f(0)$,

(2) $f \in X$ *implies* f_1, $f-f_1 \in X$, *and*

(3) $f,g \in X$ *implies* $f+g \in X$.

Then for $f \in X$,

(i) $f-f_k \in X \quad (k \in Z_+)$,

(ii) $0 \leq f(k) \leq f(0) \quad (k \in Z_+)$, *and*

(iii) $X \subset CM(Z_+)$.

Proof. For (i) write $f-f_k = (f-f_1)+(f_1-f_2)+\cdots+(f_{k-1}-f_k) \in X$ by (1) and (2). Condition (ii) follows from (i), (1), and (2).

Let $f \in X$. To show $f \in CM(Z_+)$ we must show for $n = 1,2,\cdots$ that

$$\Delta_n f(x;h_1,\cdots,h_{n-1},k) = \Delta_{n-1}f(x;h_1,\cdots,h_{n-1})$$

$$- \Delta_{n-1}f(x+k;h_1,\cdots,h_{n-1}) \geq 0$$

$(x,h_1,\cdots,h_{n-1},k \in Z_+)$. By induction, if $\Delta_{n-1}f \geq 0$ for $f \in X$, then (i) asserts $\Delta_n f = \Delta_{n-1}(f-f_k) \geq 0$, $f \in X$. □

3.6 Proposition: Let f be a real-valued function on Z_+ and define $v_n = f(n)$, $n \in Z_+$. The sequence $\{v_n\}_{n=0}^{\infty}$ is classically completely monotone if and only if f is a completely monotone function on the semigroup $(Z_+,+)$.

Proof. It is easy to check for $f \in CM(Z_+)$ that $\{v_n\}_{n=0}^{\infty}$ is a classically completely monotone sequence. Conversely, suppose $\{v_n\}_{n=0}^{\infty}$ is a classically completely monotone sequence. Thus $\Delta^k v_n \geq 0$ $(k,n \in Z_+)$. To apply Lemma 3.5, we must check that $\{v_n - v_{n+1}\}_{n=0}^{\infty}$ is still classically completely monotone. For this note

$$\Delta^k(v_n - v_{n+1}) = \Sigma_{j=0}^{k}(-1)^j\binom{k}{j}(v_{j+n} - v_{j+n+1})$$

$$= \Sigma_{j=0}^{k}(-1)^j\binom{k}{j}v_{j+n} - \Sigma_{j=1}^{k+1}(-1)^{j-1}\binom{k}{j-1}v_{j+n}$$

$$= \Sigma_{j=0}^{k+1}(-1)^j\left(\binom{k}{j} + \binom{k}{j-1}\right)v_{j+n}$$

$$= \Sigma_{j=0}^{k+1}(-1)^j\binom{k+1}{j}v_{j+n} = \Delta^{k+1}v_n \geq 0. \quad \square$$

3.7 Definition: Let $NCM(S) = \{f \in CM(S): f(1) = 1\}$, the space of normalized completely monotone functions on the commutative semi-topological semigroup S with identity 1.

3.8 Theorem: Let S be a commutative discrete semigroup with identity 1. The set $NCM(S)$ is a compact subset of $\ell^{\infty}(S)$ in

3.3.8

the weak- ($\sigma(\ell^\infty(S),\ell^1(S))$) topology (equivalently, in the pointwise topology on S). The set of extreme points* ext NCM(S) *of* NCM(S) *is precisely the set of positive semicharacters on* S, $\hat{S}_+$.

Proof. The set NCM(S) is a bounded subset of $\ell^\infty(S)$ which is closed under pointwise limits. Thus the first part of the theorem is done.

Let $f \in$ ext NCM(S). Fix $x \in S$ and write $f = (f-f_x)+f_x$. We consider three cases. Firstly suppose $f(x) = f_x(1) = 1$. Then $(f-f_x)(1) = 0$ and since $f-f_x \in CM(S)$, $f-f_x = 0$. Thus $f(xy) = f_x(y) = f(y) = 1\ f(y) = f(x)f(y)$ $(y \in S)$. Secondly, suppose $f(x) = f_x(1) = 0$. Then $f_x = 0$ since $f_x \in CM(S)$. Thus $f(xy) = f_x(y) = 0 = 0\ f(y) = f(x)f(y)$ $(y \in S)$. Thirdly, suppose $0 < f(x) < 1$. Then

$$f = (f(1)-f(x))\ \frac{f-f_x}{f(1)-f(x)} + f(x)\ \frac{f_x}{f(x)},$$

where the functions $\frac{f-f_x}{f(1)-f(x)}$ and $\frac{f_x}{f(x)}$ are in NCM(S).

Since f is an extreme point, both of these functions equal f; and thus $f(x)f = f_x$. Thus $f \in \hat{S}$ with $0 \leq f \leq 1 = f(1)$.

Now we take $f \in \hat{S}_+$ and we will argue that f must be an extreme point of NCM(S). If f is not an extreme point, then there exists $f_1, f_2 \in$ NCM(S) and λ with $0 < \lambda < 1$ so that $f = \lambda f_1+(1-\lambda)f_2$. Let μ_1, μ_2 be representing measures for f_1, f_2 respectively, supported on the compact set $c\ell$(ext NCM) $\subset \hat{S}_+$. Thus

$$f_i(x) = \int_{\hat{S}_+} \phi(x)\, d\mu_i(\phi) \qquad (i = 1,2).$$

Hence $\lambda\mu_1 + (1-\lambda)\mu_2$ is a representing measure for f supported on $\hat{S}_+$, but so is the unit point measure $\delta(f)$ with mass at $f \in \hat{S}_+$. However, representing measures are unique: the linear span of the set of functions $(x \in S)\, \phi \mapsto \phi(x) : \hat{S}_+ \to \mathbb{C}$ is uniformly dense (by the Stone-Weierstrass theorem) in the space $C(\hat{S}_+)$ of continuous functions on the compact space $\hat{S}_+$. Thus $\lambda\mu_1 + (1-\lambda)\mu_2 = \delta(f)$ implies $\mu_1 = \mu_2 = \delta(f)$. This means that $f_1 = f_2 = f$, and so f is an extreme point of NCM(S). □

3.9 Theorem: Let S be a discrete commutative semigroup with 1. Then $NCM(S) \cong M_p(\hat{S}_+)$ and $CM(S) \cong M_+(\hat{S}_+)$.

Proof. Using the argument of 3.8, each $\phi \in \hat{S}_+$ corresponds uniquely to $\delta(\phi) \in M_p(\hat{S}_+)$. But these are the extreme points of NCM(S) and $M_p(\hat{S}_+)$ respectively, and the pointwise closed convex hull of ext NCM(S) = unit ball of NCM(S) (by the Krein-Milman theorem); similarly for the unit ball of $M_p(\hat{S}_+)$ with the weak-* topology.

Finally, note that bounded pointwise convergence in CM(S) is equivalent to bounded weak-* convergence in $M_+(\hat{S}_+)$. Thus each $f \in CM(S)$ corresponds uniquely to $\mu_f \in M_+(\hat{S}_+)$ with

$$f(x) = \int_{\hat{S}_+} \phi(x)\, d\mu_f(\phi), \quad (x \in S). \quad \square$$

3.10 Corollary (Hausdorff moment theorem for Z_+). Let $\{v_n\}_{n=0}^{\infty}$ be a sequence on $S = Z_+$. There exists a unique positive measure μ on [0,1] with

$$\int_0^1 t^n d\mu(t) = v_n \quad (n \in Z_+),$$

3.3.11

if and only if $\{v_n\}_{n=0}^{\infty}$ *is classically completely monotone.*

Proof. Let $\{v_n\}_{n=0}^{\infty}$ be classically completely monotone and let $f:Z_+ \to R$ be defined by $f(n) = v_n$ $(n \in Z_+)$. By 3.6, $f \in CM(Z_+)$ and Theorem 3.9 asserts there exists a unique $\mu \in M_+(\hat{S}_+)$ with

$$f(n) = \int_{\hat{S}_+} \phi(n)\,d\mu(n), \quad (n \in Z_+).$$

But $\hat{S}_+ = \{n \mapsto t^n : t \in [0,1]\}$ (we use the convention $0^0 = 1$). So $\hat{S}_+$ is homeomorphic to $[0,1]$. For the other direction, suppose $\{v_n\}_{n=0}^{\infty}$ is given by

$$v(n) = \int_0^1 t^n d\mu(t) \quad (n \in Z_+)$$

and use Theorem 3.9. □

3.11 *Corollary: Let f be a continuous function on* $\mathbb{R}_+ = [0,\infty)$ *with* $f \in CM(\mathbb{R}_+)$. *Then there exists a unique positive measure* $\mu \in M_+([0,1])$ *with*

$$f(x) = \int_0^1 t^x d\mu(t), \quad (x \in \mathbb{R}_+).$$

Proof. Let $n \in Z_+$, and define $f_n:Z_+ \to R$ by $f_n(k) = f(k/2^n)$, $(k \in Z_+)$. The function $f_n \in CM(Z_+)$ and so there exists $\mu_n \in M_+([0,1])$ with

$$f(k/2^n) = f_n(k) = \int_0^1 t^k d\mu_n(t), \quad (k \in Z_+).$$

By a change of variables, let $\nu_n \in M_+([0,1])$ with

$$f(k/2^n) = f_n(k) = \int_0^1 t^{k/2^n} d\nu_n(t), \quad (k \in Z_+).$$

3.3.13

For each dyadic fraction r, we have

$$f(r) = \int_0^1 t^r d\nu_n(t),$$

for all sufficiently large n.

Choose a weak-* limit μ of $\{\nu_n\}_{n=1}^{\infty}$, and so

$$f(r) = \int_0^1 t^r d\mu(t), \quad (r \text{ a dyadic fraction}).$$

Since f is continuous,

$$f(x) = \int_0^1 t^x d\mu(t) \quad (x \in \mathbb{R}_+)$$

by the Lebesgue dominated convergence theorem. □

We investigate now the functions $f \in R(S)$ which are in CM(S).

3.12 Definition: Let S be a commutative semitopological semigroup with 1 and let $f \in R(S)$ have the form $f(x) = c\int_\Omega Tx d\mu$, $((T,\mu,\Omega) \in S)$ with $\mu \in M_p(\Omega)$, $c \geq 0$, and $Tx \geq 0$ $(x \in S)$. We say that the function f is completely positive, and the space of such f is denoted $R_p(S)$.

3.13 Theorem: Let S be a commutative semitopological semigroup with identity. Then $R_p(S) = CM(S)$.

Proof. Firstly, let $f \in R_p(S)$, so

$$f(x) = c\int_\Omega Tx d\mu,$$

$(T,\mu,\Omega) \in S$, $\mu \in M_p(\Omega)$, $c \geq 0$, and $Tx \geq 0$ $(x \in S)$.

Then for $x, h_1, \cdots, h_n \in S$,

3.3.14

$$\Delta_n f(x;h_1,\cdots,h_n) = c\int_\Omega Tx(\Pi_{i=1}^n (1-Th_i))\,d\mu \geq 0.$$

Thus $f \in CM(S)$.

Secondly, let $f \in CM(S)$. Thus $f \in CM(S_d)$ and by Theorem 3.9 we write (with $c = f(1) \geq 0$)

$$f(x) = c\int_{(S_d)_+^\wedge} \phi(x)\,d\mu(\phi), \quad (x \in S)$$

where $\mu \in M_p((S_d)_+^\wedge)$. Define $\hat{x}:(S_d)_+^\wedge \to R_+$ by $\hat{x}(\phi) = \phi(x)$ $(x \in S)$, and so $\hat{x} \in C((S_d)_+^\wedge) \subset L^2(\mu)$. For $x \in S$, define the operator Tx on the linear span of $\{\hat{x}:x \in S\} \subset L^2(\mu)$ by

$$Tx(\Sigma_{i=1}^n c_i\hat{x}_i) = \Sigma_{i=1}^n c_i\hat{x}\hat{x}_i \quad .$$

Since $\hat{x} \geq 0$ and $\hat{x} \leq 1$, the set $\{Tx:x \in S\}$ is a commutative set of positive contraction operators on $L^2(\mu)$. As in Theorem 1.5, it now follows that $x \mapsto Tx$ is a continuous cyclic representation of S into the unit ball of a commutative subalgebra of $(\mathcal{B}(L^2(\mu)),WO)$, namely $(L^\infty(\mu),\text{weak-}*)$; that is, $x \mapsto Tx$ is an L^∞-representation, and so $f \in \mathcal{R}_p(S)$. □

3.14 Corollary: Let S be a commutative semigroup with an identity 1 of type $\mathcal{U}$. Then $\mathcal{R}_p(S) \subset \mathcal{R}_+(S)$.

Proof. Combine Definition 1.3 and Definition 3.12. □

3.15 Corollary (Hausdorff moment theorem for $\mathbb{R}_+$): Let f be a continuous function on $\mathbb{R}_+ = [0,\infty)$. Then $f \in CM(\mathbb{R}_+)$ if and only if there exists a unique positive measure $\mu \in M_+[0,1]$ with

3.3.16

$$f(x) = \int_0^1 t^x d\mu(t), \quad (x \in \mathbb{R}_+).$$

Proof. Corollary 3.11 is the necessity. The sufficiency follows from the theorem. □

3.16 Remark: The basic results on completely monotone functions on commutative semigroups are found in Fine and Maserick [1]. The concept of moments and functions of bounded variation has been the object of much recent work. For example, see Maserick [1], Lindahl and Maserick [1], Nussbaum [1], and Newman [1] (see also Taylor [1, p. 41]). For a commutative semigroup with identity 1 and involution $x \mapsto x^*$ with a positive-definite structure (that is, for each $x \in S$, there exists a positive number L_x such that for all positive-definite functions q on S one has $q(x^*x) \leq L_x q(1)$), Saworotnow [1] has shown an abstract analogue of Bochner's theorem.

Chapter 4. Discrete separative semigroups

In this chapter we are motivated by some basic facts about representations of an infinite abelian (discrete) group G. In particular, the Fourier transforms of $\ell^1(G)$ are dense in $C(\hat{G})$ ($\hat{G}$ the compact dual group of G); the representation algebra $R(G)$ is isomorphic to $M(\hat{G})$ which is the dual space of $\ell^1(G)$ furnished with its spectral norm (the sup-norm on $\hat{G}$); the Plancherel theorem asserts that the regular representation of G, that is $\ell^1(G)$ convolving $\ell^2(G)$ (or G translating $\ell^2(G)$) is unitarily equivalent to the Fourier transforms of $\ell^1(G)$ multiplying $L^2(\hat{G})$. We will show that these situations also hold true on inverse semigroups, and in a modified way on semigroups which have enough semicharacters to separate points (called separative semigroups).

Thus the selection of material for this chapter was made to cover the above points in a reasonably complete way. We have not tried to give a survey of inverse semigroup and semicharacter theory.

The fundamental analytic object associated with a semigroup S is its semigroup algebra $\ell^1(S)$, and this is discussed in Section 1. It is shown that the maximal modular ideal space of $\ell^1(S)$ is the space $\hat{S}$ of semicharacters of S, and that $\ell^1(S)$ is semisimple if and only if $\hat{S}$ separates points on S. Section 2 is about inverse semigroups (unions of groups). Section 3 presents the important work of Hewitt and Zuckerman showing that a semigroup has enough semicharacters (separative) if and only if it is a subsemigroup of an inverse semigroup.

One of the results of Section 6 is that a semicharacter of a subsemigroup S of an inverse semigroup U can always be extended to a positive-definite function on U. Section 7 presents the construction of the regular representation of any separative semigroup, thus extending known results for groups.

§1. The semigroup algebra

1.1 Definition: The semigroup algebra $\ell^1(S)$ is the Banach space of complex functions f on S such that $||f||_1 = \Sigma_{x\in S}|f(x)| < \infty$. For $x \in S$ define $\delta_x \in \ell^1(S)$ by $\delta_x(x) = 1$, $\delta_x(y) = 0$ $(y \in S,\ y \neq x)$, then each $f \in \ell^1(S)$ may be written $f = \Sigma_{x\in S} f(x)\delta_x$. Let $c_c(S)$ be the space of finite linear combinations of $\{\delta_x : x \in S\}$.

1.2 Proposition: $\ell^1(S)$ is a commutative Banach algebra, whose maximal (modular) ideal space may be identified with S.

Proof. Given $f,g \in \ell^1(S)$, $x \in S$, define $f*g(x) = \Sigma\{f(y)g(z) : y,z \in S,\ yz = x\}$, an absolutely convergent sum. Further

$$\Sigma_x|f*g(x)| \leq \Sigma_{x\in S}\Sigma_{yz=x}|f(y)||g(z)| = ||f||_1||g||_1.$$

Associativity follows from the associativity of S and the fact that $c_c(S)$ is norm-dense in $\ell^1(S)$.

Let $\chi \in \hat{S}$, that is $\chi(xy) = \chi(x)\chi(y)$ $(x,y \in S)$, $|\chi(x)| \leq 1$ and $\chi \not\equiv 0$. For $f \in \ell^1(S)$ let $\chi^\#(f) = \Sigma_{x\in S} f(x)\chi(x)$, (absolutely convergent) so $|\chi^{\#}(f)| \le ||f||_1$. Further $\chi^{\#}$ is multiplicative on $c_c(S)$, thus $\chi^{\#}$ is a bounded multiplicative linear functional. Conversely if ϕ is a bounded multiplicative linear functional on $\ell^1(S)$ then $\chi : x \mapsto \phi(\delta_x)$ is a semicharacter of S and $\phi = \chi^{\#}$. □

1.3 <u>Definition</u>: $\hat{S}$ is a locally compact space with the Gelfand topology induced by $\ell^1(S)$, equivalent to the topology of pointwise convergence on S, and for $f \in \ell^1(S)$ let $||f||_{sp} = \sup_{\chi \in \hat{S}} |\chi^{\#}(f)|$, the <u>spectral norm</u> of $\ell^1(S)$. Note that $||f||_{sp} = \lim_{n\to\infty} ||f^n||_1^{1/n}$ where f^n is inductively defined by $f^n = f^{n-1} * f$. Define the <u>Fourier transform</u> of $f \in \ell^1(S)$ by $\hat{f}(\chi) = \Sigma_x f(x)\chi(x)$, $(\chi \in \hat{S})$, then $\hat{f} \in c_o(\hat{S})$.

1.4 <u>Proposition</u>: <u>$\ell^1(S)$ is semisimple if and only if $\hat{S}$ separates S, that is, for each $x,y \in S$ with $x \neq y$ there exists $\chi \in \hat{S}$ such that $\chi(x) \neq \chi(y)$.</u>

Proof. If $\ell^1(S)$ is semisimple, then for $x,y \in S$, $x \neq y$, $\delta_x - \delta_y \in \ell^1(S)$ and is not zero, so there exists $\chi \in \hat{S}$ such that $\chi(x) - \chi(y) \neq 0$. Conversely, if $\hat{S}$ separates points, then the span of $\hat{S}$ is a self-adjoint algebra of bounded functions separating the points of S (note: $\chi \in \hat{S}$ implies $\overline{\chi} \in \hat{S}$, and $1 \in \hat{S}$). Let A be the sup-norm (over S) closure of $sp(\hat{S})$, then by the Stone-Weierstrass theorem $A \cong C(X)$, X some compact Hausdorff space. Since $\hat{S}$ separates S there exists a one-to-one map $j: S \to X$, and thus an induced map $j': \ell^1(S) \to M(X)$ such that $\int_X g\,d(j'f) = \Sigma_{x\in S} f(x) g(jx)$, $(f \in \ell^1(S),\ g \in C(X))$. But $\Sigma_{x\in S} f(x)\chi(x) = 0$ for all $\chi \in \hat{S}$ implies $j'f = 0$, and j' is one-to-one, so $\ell^1(S)$ is semisimple. □

§2. Inverse semigroups

2.1 Notation: Let U denote an inverse semigroup, that is, a semigroup which is a union of groups. For $x \in U$, let $H(x)$ denote the maximal group containing x, e_x the identity in $H(x)$, and x' the inverse of x in $H(x)$.

Let $E(U)$ denote the set of idempotents in U, thus $E(U)$ is a subsemigroup of U. The map $x \mapsto e_x$ is a homomorphism of U onto $E(U)$. Note that $E(U)$ is a lower semilattice, where $e_1 \geq e_2$ means $e_1e_2 = e_2$ and $e_1 \wedge e_2 = e_1e_2$ $(e_1, e_2 \in E(U))$. Further if $e_1 \geq e_2$ then the map $x \mapsto e_2x$ restricted to $H(e_1)$ is a group-homomorphism of $H(e_1)$ into $H(e_2)$ (let $x \in H(e_1)$, then $(xe_2)(x'e_2) = e_1e_2 = e_2$, and $(xe_2)e_2 = xe_2$, thus $xe_2 \in H(e_2)$ and $(xe_2)' = x'e_2$). This fact will be of use in extending characters of $H(e)$ to semicharacters of U.

We recall a standard theorem from harmonic analysis:

2.2 Theorem: Let G be a discrete abelian group, then the dual of G, the group $\hat{G}$ of characters of G, is a compact abelian group and separates the points of G. Further $\hat{G}$ has a unique translation-invariant probability measure $m_{\hat{G}}$ (Haar) and the Fourier transform extends to an isometry of $\ell^2(G)$ onto $L^2(\hat{G}, m_G)$ (the Plancherel theorem).

2.3 Theorem: Let U be an inverse semigroup, then each $\chi \in U$ satisfies $|\chi(x)| = 0$ or 1, and $\chi(x') = \overline{\chi(x)}$ $(x \in U)$, and U is a union of groups with $\chi' = \overline{\chi}$ $(\chi \in \hat{U})$. Unless U has an identity it may happen that $\chi_1\chi_2 = 0$ for some $\chi_1, \chi_2 \in \hat{U}$. In any case $\hat{U} \cup \{0\}$ is a semigroup.

Proof. Let $\chi \in \hat{U}$, $x \in U$. Then $\chi(x) = \chi(x)\chi(e_x)$ and $\chi(x)\chi(x') = \chi(e_x)$. If $\chi(x) \neq 0$, then $\chi(e_x) = 1$, and so $\chi(x')\chi(x) = 1$. But $|\chi(x)| \leq 1$ and $|\chi(x')| \leq 1$ implying that $|\chi(x)| = 1$ and $\chi(x') = \overline{\chi(x)}$. The set $\{\phi \in \hat{U}: |\phi| = |\chi|\}$ is a locally compact subgroup of $\hat{U}$ with identity $|\chi|$, and $\hat{U}$ is the union of such groups. The product of two semicharacters is a semicharacter unless it is zero. □

2.4 Theorem: Let H(e) be a maximal group in U, then $\hat{U}$ contains an isomorphic image of $H(e)^\wedge$ denoted by $\Gamma(e)$. The identity π_e of $\Gamma(e)$ is the semicharacter defined by $\pi_e(x) = 1$ if $e_x \geq e$ and $\pi_e(x) = 0$ otherwise, that is, $e_x e \neq e$. The group $\Gamma(e)$ is the maximal group containing π_e.

Proof. Let $\gamma \in H(e)^\wedge$ and define $\gamma^b \in \hat{U}$ by

$$\gamma^b(x) = \begin{cases} \gamma(ex) & e_x \geq e \\ 0 & e_x e \neq e \end{cases}.$$

We show that γ^b is a semicharacter. Let $y, z \in U$ with $e_y \geq e$ and $e_z \geq e$, then $e_{yz} \geq e$ and $\gamma^b(yz) = \gamma(eyz) = \gamma(eyez)$ $= \gamma(ey)\gamma(ez) = \gamma^b(y)\gamma^b(z)$. Further γ^b is zero on $\{x: e_x e \neq e\}$, an ideal in U, so γ^b is multiplicative on U. Since $H(e)^\wedge$ is determined by its values on $H(e)$ we see that $H(e)^\wedge$ is mapped isomorphically into $\hat{U}$. If $\chi \in \hat{U}$ with $|\chi| = \pi_e$ then $\chi(x) = 0$ for $e_x e \neq e$, and $\chi(x) = \chi(ex)$ for $e_x e = e$, but $\chi|H(e)$ is in $\hat{H}(e)$. □

2.5 Corollary: If U is an inverse semigroup then $\hat{U}$ separates the points of U and thus $\ell^1(U)$ is semisimple.

Proof. Let $x, y \in U$ with $x \neq y$. If $y \in H(x)$ then there exists $\gamma \in H(x)^\wedge$ with $\gamma(x) \neq \gamma(y)$. Thus $\gamma^b \in \hat{U}$ and $\gamma^b(x) \neq \gamma^b(y)$. If $y \notin H(x)$ then $e_x e_y \neq e_y$ or $e_x e_y \neq e_x$. Suppose $e_x e_y \neq e_y$, then there exists $\pi \in \hat{U}$ with $\pi(y) = 1$ and $\pi(x) = 0$. □

2.6 Theorem: Suppose U is a finite union of groups $H(e_1), \cdots, H(e_n)$, then $\hat{U}$ is the union of $\Gamma(e_j)$, $j = 1, \cdots, n$.

Proof. From Theorem 2.4 it suffices to show that $\pi_{e_1}, \cdots, \pi_{e_n}$ are the only idempotents in $\hat{U}$. Let π be an idempotent in $\hat{U}$ and let $F = \{e \in E(U) : \pi(e) = 1\}$. By hypothesis F is finite so F contains a minimal element e_j some j (the product of all elements of F). Thus $\pi = \pi_{e_j}$. □

We will need the above theorem in later work dealing with the regular representation.

2.7 Definition: Let U be an inverse semigroup and let $f \in \ell^1(U)$. Define $f^*(x) = \overline{f(x')}$ $(x \in U)$. With this involution $\ell^1(U)$ is a symmetric Banach $*$-algebra, $(f^*)^\wedge = \bar{\hat{f}}$, and the Šilov boundary of $\ell^1(U)$ is all of $\hat{U}$, (standard Banach algebra theory).

§3. Separative semigroups

We observe that if S is any subsemigroup of an inverse semigroup U then $\hat{S}$ separates S, since $\hat{S}$ contains the restrictions of $\hat{U}$ to S. If S is a semigroup for which $\hat{S}$ separates S then S has the property $x^2 = xy = y^2$ implies $x = y$ $(x,y \in S)$. Indeed suppose $x^2 = xy = y^2$ and $\chi \in \hat{S}$, then $\chi(x)^2 = \chi(x)\chi(y) = \chi(y)^2$, implying $\chi(x)[\chi(x)-\chi(y)] = 0$ and $\chi(y)[\chi(x)-\chi(y)] = 0$, that is $\chi(x) = \chi(y)$. The remarkable fact discovered by Hewitt and Zuckerman [2], is that this property characterizes subsemigroups of inverse semigroups. We will prove the following fundamental theorem in several stages.

3.1 Theorem: *Let S be a discrete semigroup then the following are equivalent*:

1) $\hat{S}$ *separates the points of* S;
2) S *is a subsemigroup of an inverse semigroup*;
3) $x^2 = xy = y^2$ *implies* $x = y$ $(x,y \in S)$.

Such a semigroup will be called *separative*.

We have already shown 2) implies 1), 1) implies 3). To show 3) implies 2) we will define an equivalence relation on S for which each equivalence class is a semigroup with the cancellation property, and thus each class can be embedded in a group.

3.2 Definition: For $x \in S$ define $h_x = \{y \in S : y^n = ux,\ x^m = vy$ for some $u,v \in S,\ m,n = 1,2,\cdots\}$.

4.3.5

3.3 Proposition: The relation $x \sim y$ if and only if $y \in h_x$ is an equivalence relation on S.

Proof. Clearly $x \sim x$, and $x \sim y$ implies $y \sim x$. Let $y \in h_x$, $z \in h_y$ $(x,y,z \in S)$. Then there exist $u,v \in S$, $m,n = 1,2,3,\cdots$, such that $y^n = ux$, $z^m = vy$ thus $z^{mn} = v^n y^n = v^n ux$. Similarly one can find $w \in S$, $k = 1,2,\cdots$, such that $x^k = wz$, and thus $z \in h_x$. □

The sets $\{h_x\}$ are called the archimedean components of S.

3.4 Proposition: Each h_x is a subsemigroup of S, and if $u \in h_x$, $v \in h_y$ then $uv \in h_{xy}$ $(u,v,x,y \in S)$. Thus the set $E = \{h_x : x \in S\}$ with the multiplication $h_x \cdot h_y = h_{xy}$ is a semigroup. Since $x^2 \in h_x$, E is an idempotent semigroup.

Proof. Let $u \in h_x$, $v \in h_y$, then $u^m = wx$, $v^n = zy$, some $w,z \in S$, m,n positive integers. Suppose $m \geq n$, then $(uv)^m = wxv^{m-n}\ zy = w'xy$, some $w' \in S$. Similarly there exist $w'' \in S$, $k = 1,2,\cdots$, with $(xy)^k = w''uv$. Thus $uv \in h_{xy}$. Note $x^2 \in h_x$, so $u,v \in h_x$ implies $uv \in h_{xx} = h_x$, thus h_x is a subsemigroup of S. □

Warning: Multiplying the h_x's as subsets of S yields only that $h_x h_y \subset h_{xy}$. Equality may fail to hold. It is easy to see that if $y \in h_x$ $(y,x \in S)$ then $\chi(x) = 0$ implies $\chi(y) = 0$ $(\chi \in \hat{S})$.

3.5 Theorem: Let S be a semigroup with the property

3.1.(3): $x^2 = xy = y^2$ *implies* $x = y$ $(x,y \in S)$,

then each h_x *has the cancellation property, that is,* $y,z \in h_x$, $yx = zx$ *implies* $y = z$.

Proof. Property 3.1.(3) implies that $x^n y = x^n z$ implies $xy = xz$ $(x,y,z \in S,\ n = 2,3,\cdots)$. Indeed one may assume that $n = 2m$ (m integer) (else multiply by x) then $(x^m y)^2 = x^n y^2 = x^n yz = (x^m y)(x^m z) = x^n z^2 = (x^m z)^2$ so $x^m y = x^m z$. Now $m \leq (n+1)/2$, so this process leads to $m = 1$ in finitely many steps.

Let $y \in h_x$ so $y^n = ux$, some $u \in S$, $n = 1,2,\cdots$, thus $yu \in h_y h_u = h_x h_u \subset h_{xu} = h_{y^n} = h_y = h_x$, thus we can write $y^{n+1} = (yu)x$ with $yu \in h_x$.

Finally let $y,z \in h_x$ with $yx = zx$. Then there exist $u,v \in h_x$, integers m,n with $y^n = ux$, $z^m = vx$. Then $y^{n+1} = uxy = uxz = y^n z$, implying $y^2 = yz$ (from above remarks). Similarly $z^{m+1} = vxz = vxy = z^m y$, implying $z^2 = zy$. Finally property 3.1.(3), applied again, shows $y = z$. □

We continue to require that S has property 3.1.(3).

3.6 Definition: Let S_1 be the subsemigroup of $S \times S$ defined by $S_1 = \{(x_1,x_2) \in S \times S : x_2 \in h_{x_1}\}$. Proposition 3.4 shows S_1 is indeed a semigroup.

3.7 Proposition: *The relation* $\sim$ *defined on* S_1 *by* $(x_1,x_2) \sim (y_1,y_2)$ *if and only if* $x_1 y_2 = x_2 y_1$ *and* $y_1 \in h_{x_1}$ *is an equivalence relation. Further* $(x_1,x_2) \sim (y_1,y_2)$ *and* $(u_1,u_2) \sim (v_1,v_2)$ *implies* $(x_1 u_1, x_2 u_2) \sim (y_1 v_1, y_2 v_2)$, *and so* $S_1/\sim$ *is a semigroup.*

Proof. The relation $\sim$ is clearly reflexive and symmetric. Let $(x_1,x_2)\sim(y_1,y_2)$ and $(y_1,y_2)\sim(z_1,z_2)$, then $x_1y_2 = x_2y_1$, $y_1z_2 = y_2z_1$, and all six points lie in h_{x_1}. One obtains $x_2y_1z_1 = x_1y_2z_1 = x_1y_1z_2$, and cancellation in $h_{x_1} = h_{y_1}$ implies $x_2z_1 = x_1z_2$. Thus $\sim$ is transitive. For the second statement in the theorem one multiplies the equation $x_1y_2 = x_2y_1$ with $u_1v_2 = u_2v_1$ to obtain $x_1u_1y_2v_2 = x_2u_2v_1y_1$. Also $y_1 \in h_{x_1}$, $v_1 \in h_{u_1}$ imply $y_1v_1 \in h_{x_1u_1}$ (Proposition 3.4), thus $(x_1u_1,x_2u_2)\sim(y_1v_1,y_2\ v_2)$. □

3.8 <u>Definition</u>: Denote the equivalence class under $\sim$ of $(x_1,x_2) \in S_1$ by $[x_1,x_2]$ and let U denote the semigroup $S_1/\sim$. For $x \in S$ let $H(x)$ denote the image of $h_x \times h_x$ in U. (We will show this agrees with earlier notation.)

3.9 <u>Theorem</u>: <u>U is an inverse semigroup and S is isomorphic to a subsemigroup of U. The maximal groups in U are exactly the sets</u> $H(x)$, $x \in S$.

Proof. It is clear that for $x,y \in S$ that either $H(x) = H(y)$ or $H(x) \cap H(y) = \emptyset$ (since either $h_x = h_y$ or $h_x \cap h_y = \emptyset$), and that U is the union of the $H(x)$'s.

Let $x \in S$, then $H(x)$ is a group with identity $[x,x]$. Indeed let $u,v \in h_x$, then $(xu,xv)\sim(u,v)$ so that $[x,x][u,v] = [xu,xv] = [u,v]$; and $[u,v][v,u] = [uv,uv] = [x,x]$ since $uv \in h_x$. Define the map $\pi:S \to U$ by $\pi x = [x^2,x]$ $(x \in S)$, then $\pi x \in H(x)$ (since

4.3.10

$x^2 \in h_x$). For $x,y \in S$, $\pi(xy) = [x^2y^2, xy] = [x^2,x][y^2,y]$ $= (\pi x)(\pi y)$. Suppose $\pi x = \pi y$, then $H(x) = H(y)$ so $y \in h_x$ and $(x^2,x) \sim (y^2,y)$, implying $x^2y = xy^2$. But by Theorem 3.5, this implies $xy = y^2$ and thus $x = y$.

Since $H(x) \cap H(y) = \emptyset$ implies $[x,x] \neq [y,y]$ and each $H(x)$ is a group we see that $H(x) = H([x,x])$ (the maximal group containing $[x,x]$). □

This concludes the proof of Theorem 3.1.

3.10 __Proposition__: __The restrictions of__ $\hat{U}$ __to__ S __are exactly the elements of__ $\hat{S}$ __which satisfy__ $|\chi(x)| = 0$ __or__ 1 $(x \in S)$.

Proof. Theorem 2.3 shows $|\chi(x)| = 0$ or 1 $(x \in S, \chi \in \hat{U})$. Let $\phi \in \hat{S}$ with $|\phi(x)| = 0$ or 1 $(x \in S)$. Define a function χ on U by

$$\chi([x,y]) = \begin{cases} \phi(x)/\phi(y) & \text{if } \phi(y) \neq 0 \\ 0 & \text{if } \phi(y) = 0, \ (x \in S, y \in h_x). \end{cases}$$

This is well-defined, for if $(x,y) \sim (u,v)$ then $xv = uy$ so $\phi(x)\phi(v) = \phi(y)\phi(u)$, and if $\phi(y) \neq 0$ then $\phi(v) \neq 0$ and $\phi(x)/\phi(y) = \phi(u)/\phi(v)$. Note that $v \in h_y$ so there exists $w \in S$, an integer m such that $v^m = wy$ and so $\phi(v)^m = \phi(w)\phi(y)$. Thus $\phi(y) = 0$ implies $\phi(v) = 0$, and $\chi([x,y]) = 0$. Also if $\chi([x,y]) \neq 0$, then $|\chi([x,y])| = 1$ and thus χ is a bounded function on U. It agrees with ϕ on the image of S in U since $\chi([x^2,x]) = \phi(x)$, $(x \in S)$. To show χ is a semicharacter consider two points $[x,y]$, $[u,v] \in U$. If $\phi(yv) \neq 0$ then

$\chi([x,y][u,v]) = \chi([xu,yv]) = \phi(xu)/\phi(yv) = (\phi(x)/\phi(y))(\phi(u)/\phi(v))$ $= \chi([x,y])\chi([u,v])$. If $\phi(yv) = 0$ then one of $\phi(y),\phi(v)$ is 0, so one of $\chi([x,y]),\chi([u,v]) = 0$. □

3.11 Proposition: If $x \in S$ has an inverse $x' \in S$, such that $e = xx'$ is an idempotent and $ex = x$, then h_x is a group, $x',e \in h_x$ and h_x may be identified with $H(x) \subset U$.

Proof. One may assume that $ex' = x'$, so $(xx')x' = x'$ and also $(xx')x = x$ implying $x' \in h_x$ and $e \in h_x$. For each $y \in h_x = h_e$ one has $ye^2 = ye$ but cancellation holds in h_x so $ye = y$. Further there exists $v \in h_x$ such that $yv = e$ (definition of h_e, and trick from Theorem 3.5). So h_x is a group. Let $[u,v] \in H(x)$, and let $y = uv'$ (v' the inverse of v in h_x), then $[u,v] = [y^2,y] = \pi y$, thus h_x is identified with $H(x) \subset U$. □

3.12 Corollary: If $x \in S$ is such that h_x contains an idempotent e, then h_x is a group. (Note $h_x = h_e$ and e has an inverse.)

We will later show that if S is proper in U then the restrictions of $\hat{U}$ to S are proper in $\hat{S}$. That is, there exist $\chi \in \hat{S}$, $x \in S$ such that $0 < |\chi(x)| < 1$. This will be proved by Banach algebra techniques applied to $\ell^1(S)$.

§4. The Šilov boundary of $\ell^1(S)$

In this section, S will be a separative semigroup canonically embedded into an inverse semigroup U, as constructed in Section 3. Recall from Section 1 that the maximal (modular) ideal space of

$\ell^1(S)$ is identified with $\hat{S}$. Theorem 3.10 shows that the semicharacters of U are determined by their action on S, so we can consider $\hat{U}$ as a subset of $\hat{S}$.

4.1 Theorem: The Šilov boundary of $\ell^1(S)$ is a subset of $\hat{U}$ (considered as a subset of $\hat{S}$), that is, for each $f \in \ell^1(S)$ there exists $X \in \hat{U}$ such that $|\hat{f}(X)| = ||f||_{sp}$.

Proof. Observe that $\ell^1(S)$ is a closed subalgebra of $\ell^1(U)$, so the spectral norm of $f \in \ell^1(S)$ is the same in both algebras. But $\hat{U}$ is the Šilov boundary of $\ell^1(U)$ so there exists $X \in \hat{U}$ such that $|\hat{f}(X)| = ||f||_{sp}$. Further note that $\hat{U}$ is closed in $\hat{S}$ since by 3.10 $\hat{U}$ is closed under pointwise limits. □

4.2 Example: The Šilov boundary of $\ell^1(S)$ may be a proper subset of $\hat{U} \subset \hat{S}$. Indeed let S be the additive semigroup $Z_+ = \{0,1,2,\cdots\}$. Then $h_0 = \{0\}$ and $h_1 = \{1,2,\cdots\}$. The construction of Section 3 yields $H(0) = \{0\}$ and $H(1) \cong Z$. Denote the zero-element of H(1) by 0'. The semicharacters of U are ϕ and $\{X_\theta : 0 \leq \theta < 2\pi\}$ defined by $\phi(0) = 1$, $\phi(H(1)) = 0$ and $X_\theta(0) = 1$, $X_\theta(0') = 1$, $X_\theta(n) = e^{in\theta}(n \in H(1))$. Let $f \in \ell^1(S)$, so $f = \Sigma_{n=0}^{\infty} a_n \delta_n$, $\Sigma_n |a_n| < \infty$. We have $\hat{f}(\phi) = a_0$ and $\hat{f}(X_\theta) = a_0 + \sum_{n=1}^{\infty} a_n e^{in\theta}$. However $\hat{f}(\phi) = a_0 = \frac{1}{2\pi}\int_0^{2\pi}\hat{f}(X_\theta)\,d\theta$, thus ϕ is not in the Šilov boundary, which is $\{X_\theta : 0 \leq \theta < 2\pi\}$.

4.3 Theorem: Let $x \in S$ such that x has no inverse in S, then there exists $X \in \hat{S}$ with $0 < |X(x)| < 1$, indeed $0 < |X(y)| < 1$ for all $y \in h_x$.

4.4.3

Proof. By Corollary 3.12 h_x is not a group and has no idempotent. Thus h_x is identified with a subsemigroup of H(x) which does not contain the identity. Consider $\ell^1(h_x)$ as a subalgebra of $\ell^1(H(x))$. We will show that $H(x)^\wedge|h_x \neq \hat{h}_x$. Suppose not, then the maximal modular ideal space of $\ell^1(h_x)$ is $H(x)^\wedge|h_x$. But $|\hat{\delta}_x| = 1$ on $H(x)^\wedge|h_x$ which implies that δ_x has an inverse f in $\ell^1(h_x)$ (the function $F(z) = 1/z$ operates on δ_x). Then $\hat{\delta}_x\hat{f} = 1$ on $H(x)^\wedge$, but $\ell^1(H(x))$ is semisimple and the unique inverse of δ_x is $\delta_{x'}$, so $f = \delta_{x'}$. This is a contradiction to x having no inverse.

Hence there exists $\chi \in \hat{h}_x$ such that $|\chi(x)| < 1$. Further $0 < |\chi(y)| < 1$ for all $y \in h_x$. Indeed let $y \in h_x$, then $y^n = ux$, some $u \in h_x$, integer n, so that $|\chi(y)|^n = |\chi(u)||\chi(x)| \leq |\chi(x)| < 1$. If $\chi(x) = 0$, then $\chi(y) = 0$ for all $y \in h_x$ contrary to $\chi \not\equiv 0$, thus $\chi(x) \neq 0$. Similarly $\chi(y) = 0$ for some $y \in h_x$ implies $\chi(x) = 0$ (since $h_x = h_y$), thus $0 < |\chi(y)| < 1$ for all $y \in h_x$.

We now extend χ to be a semicharacter χ^b on all of S by defining

$$\chi^b(z) = \begin{cases} \chi(zx)/\chi(x) & \text{for } zx \in h_x, \\ 0 & zx \notin h_x. \end{cases}$$

We first show that χ^b is multiplicative. The set $\{z \in S : zx \notin h_x\}$ is an ideal in S, so it suffices to consider y,z such that $yx, zx \in h_x$. Then

4.5.1

$$\chi^b(y)\chi^b(z) = \frac{\chi(yx)}{\chi(x)}\,\frac{\chi(zx)}{\chi(x)} = \frac{\chi(yzx^2)}{\chi(x)\chi(x)} = \chi(yzx)/\chi(x) = \chi^b(yz)$$

(since $yzx \in h_x$). Also $\chi^b|h_x = \chi$. To show χ^b is bounded let $y \in S$ with $yx \in h_x$. Then $y^n x \in h_{y^n}h_x = h_y h_x \subset h_{xy} = h_x$ for each $n = 1,2,\cdots$, and $\chi(y^n x) = \chi^b(y^n)\chi(x) = \chi^b(y)^n\chi(x)$, which implies that $|\chi^b(y)|^n \leq |\chi(y^n x)/\chi(x)| \leq |1/\chi(x)|$ for all n, so $|\chi^b(y)| \leq 1$. Thus $\chi^b \in \hat{S}$ and $0 < |\chi^b(y)| = |\chi(y)| < 1$ $(y \in h_x)$. □

§5. The representation algebra

In this section S denotes a separative semigroup, and U denotes the inverse semigroup containing S, as in Theorem 3.9.

5.1 <u>Theorem</u>: <u>Let</u> S <u>be a separative semigroup then</u> $R(S)$ <u>is isometrically isomorphic to the dual space of</u> $\ell^1(S)$ <u>furnished with the spectral norm. Thus</u> $R(S)$ <u>is the dual of a function algebra. The isomorphism is given by</u> $\phi_0(x) = \phi(\delta_x)$ $(\phi_0 \in R(S), x \in S$, ϕ <u>in the dual</u> of $(\ell^1(S), ||\cdot||_{sp}))$.

Proof. The idea is to identify the dual of $(\ell^1(S), ||\cdot||_{sp})$ with a quotient space of the space of finite regular Borel measures $M(\hat{U})$ on the locally compact Hausdorff space $\hat{U}$. Note that $(\ell^1(S), ||\cdot||_{sp})$ is isomorphic to a subalgebra of $C_o(\hat{U})$, and indeed $(\ell^1(U), ||\cdot||_{sp})$ is dense in $C_o(\hat{U})$.

Let ϕ be a $||\cdot||_{sp}$-bounded linear functional, then by the Hahn-Banach and Riesz theorems there exists $\mu \in M(\hat{U})$ such that $\phi(f) = \int_{\hat{U}}\hat{f}d\mu$ $(f \in \ell^1(S))$, and $||\phi|| = ||\mu||$.

4.5.3

Write $d\mu = g d\nu$ where ν is a probability measure on $\hat{U}$ and $g \in L^1(\nu)$. Then $\phi_0(x) = \phi(\delta_x) = \int_{\hat{U}} \hat{x} g d\nu$, $(x \in S)$, but $x \mapsto \hat{x}$ is an L^∞-representation in $(\nu, \hat{U})$. Thus $\phi_0 \in R(S)$ and $||\phi_0|| \leq ||g||_1 = ||\mu|| = ||\phi||$.

Conversely, let $f \in R(S)$ and $\varepsilon > 0$, then there exists an L^∞-representation (T, μ, Ω) of S and $g \in L^1(\mu)$ such that $f(x) = \int_\Omega Txg d\mu$ and $||g||_1 < ||f||_R + \varepsilon$.

Define the bounded linear map $T_1 : \ell^1(S) \to L^\infty(\mu)$ by $T_1(\Sigma_x h(x)\delta_x) = \Sigma_x h(x) Tx$, then $||T_1 h||_\infty \leq ||h||_1$, and $T_1(h*k) = (T_1 h)(T_1 k)$ $(h,k \in \ell^1(S))$. Thus $T_1(h^n) = (T_1 h)^n$ and $||(T_1 h)||_\infty^n = ||(T_1 h)^n||_\infty = ||T_1(h^n)||_\infty \leq ||h^n||_1$, implying that $||T_1 h||_\infty \leq ||h^n||_1^{1/n}$, all n. Let $n \to \infty$, then we have $||T_1 h||_\infty \leq ||h||_{sp}$ $(h \in \ell^1(S))$ (see 1.3).

Finally define the bounded linear functional ϕ on $\ell^1(S)$ by $\phi(\Sigma_x h(x)\delta_x) = \Sigma_x h(x) f(x) = \int_\Omega T_1(\Sigma h(x)\delta_x) g d\mu$, and note that $|\phi(h)| \leq ||h||_{sp} ||g||_1 \leq ||h||_{sp}(||f||_R + \varepsilon)$; also ε is arbitrary and $f = \phi_0$, $(f(x) = \phi(\delta_x)$, $x \in S)$. □

5.2 <u>Corollary</u>: *If U is an inverse semigroup then $R(U) \cong M(\hat{U})$. Multiplication in $R(U)$ corresponds to convolution of measures on $\hat{U}$ (which is almost a topological semigroup, except that the convolution of two point masses may be zero).*

5.3 <u>Theorem</u>: *Let S be a separative semigroup and let S_o be a subsemigroup of S, then $R(S)|S_o = R(S_o)$.*

Proof. Associate an inverse semigroup U to S as above. The completions of $(\ell^1(S_o), ||\cdot||_{sp})$ and $(\ell^1(S), ||\cdot||_{sp})$ will be identified with closed subalgebras A_o and A of $C_o(\hat{U})$, respectively. Thus by Theorem 5.1, $R(S_o) \cong A_o^*$ and $R(S) \cong A^*$. The injection $A_o \to A$ is isometric, therefore the adjoint, $A^* \to A_o^*$, which is the restriction map, is onto (Hahn-Banach theorem). □

5.4 Theorem: Let U be an inverse semigroup, then the set of positive-definite functions on U is in one-to-one correspondence with the set of positive measures in $M(\hat{U})$, under the relation $f(x) = \int_{\hat{U}} \hat{x} d\mu$ $(x \in S)$.

Proof. Let $\mu \in M(\hat{U})$ with $\mu \geq 0$, let $f(x) = \int_{\hat{U}} \hat{x} d\mu$, $(x \in U)$ and choose points $x_1, \cdots, x_n \in U$, numbers $c_1, \cdots, c_n \in \mathbb{C}$. Then

$\Sigma_{i,j=1}^n c_i \bar{c}_j f(x_i x_j') = \int_{\hat{U}} \Sigma_{i,j} c_i \bar{c}_j \hat{x}_i (x_j')\hat{} d\mu = \int_{\hat{U}} |\Sigma c_i \hat{x}_i|^2 d\mu$

(since $(x')\hat{} = \bar{\hat{x}}$, $x \in U$)

$\geq |\int_{\hat{U}} \Sigma c_i \hat{x}_i d\mu|^2 / (\mu \hat{U}) = |\Sigma_{i=1}^n c_i f(x_i)|^2 / (\mu \hat{U})$.

Thus f is positive-definite.

Conversely let f be positive-definite on U, then $f \in R(U)$ by Theorem 3.1.5. By Theorem 5.1 there exists $\mu \in M(\hat{U})$ with $||\mu|| = ||f||_R$ and $f(x) = \int_{\hat{U}} \hat{x} d\mu$. We first consider the case where U has an identity 1. Then $||f||_R = f(1) = \int_{\hat{U}} 1 \, d\mu$, and thus $\mu \geq 0$ (since $||f||_R = ||\mu||$).

Now suppose U has no identity, so let U_1 denote $U \cup \{1\}$, then $\hat{U}_1$ is $\hat{U} \cup \{\chi_o\}$, where $\chi_o(1) = 1$, $\chi_o(U) = 0$. The hypotheses imply that f extends to be positive-definite on U_1, and thus there exists a positive measure μ_1 on U_1 such that $f(x) = \int_{\hat{U}_1} \hat{x} d\mu_1$. Let $\mu = \mu_1 | \hat{U}$, then $f(x) = \int_{\hat{U}} \hat{x} d\mu$ for $x \in U$, since $\hat{x}(\chi_o) = 0$, and so μ is the required measure. □

This theorem is the version of the Bochner theorem which holds for inverse semigroups.

Since there are no continuity problems on discrete semigroups, we can give a version of the Bochner-Eberlein theorem for any separative semigroup.

5.5 Theorem: Let S be a separative semigroup and let $\{f_\alpha\}$ be a net in $R(S)$ which converges pointwise on S to some function f. If $||f_\alpha||_R \leq M < \infty$ for all α, then $f \in R(S)$ and $||f||_R \leq M$.

Proof. $\{f_\alpha\}$ is a net in a bounded set in $R(S)$, the dual space of $(\ell^1(S), ||\cdot||_{sp})$, hence by the Alaoglu theorem, $\{f_\alpha\}$ has a weak-* cluster point $g \in R(S)$, with $||g||_R \leq M$. For $x \in S$, pairing f_α with δ_x shows that $g(x)$ is a cluster point of $\{f_\alpha(x)\}$, thus $f = g \in R(S)$. □

5.6 Remark: We may combine Theorem 5.1 with Theorem 3.2.11 to obtain a characterization of $R(S)$ for S being a semitopological semigroup of type U. Let V be the dense inverse semigroup in S, and take V discrete. Theorem 3.2.11 shows $R(S) = C^B(S) \cap R(V)$, with identical R-norms. Thus $f \in C^B(S)$ is in $R(S)$ if and only if $|\Sigma_{i=1}^n c_i f(x_i)| \leq K_f ||\Sigma_{i=1}^n c_i \delta_{x_i}||_{sp}$ for each $n = 1,2,\cdots$, $c_i \in \mathbb{C}$, $x_i \in V$, $(1 \leq i \leq n)$, some constant $K_f < \infty$. Note that $||f||_R$ is the least value of K_f that works.

§6. Extensions of semicharacters

The problem treated in this section is the possibility of extending semicharacters of a semigroup to semicharacters of a larger semigroup. We will show this can be done for semicharacters χ which satisfy $|\chi| = 1$. For the case of a subsemigroup of an inverse semigroup we will extend semicharacters to positive-definite functions.

6.1 Theorem: Let S be a separative semigroup and let S_o be a subsemigroup of S. Suppose $\chi \in \hat{S}_o$ and $|\chi| = 1$ then there exists $\psi \in \hat{S}$ such that $\psi|S_o = \chi$.

Proof. Embed S canonically into an inverse semigroup U (as in Theorem 3.9). Let U_o be the inverse subsemigroup of U which is generated by S_o, that is, U_o is the set of products $\{xy' : x,y \in S_o\}$ (note $x = x^2x'$). Define $\chi_1 \in \hat{U}_o$ by

$$\chi_1(xy') = \chi(x)/\chi(y) \qquad (x,y \in S_o).$$

Clearly $|\chi_1| = 1$ and χ_1 is well-defined, for if $x_1y_1' = x_2y_2'$ $(x_1,x_2,y_1,y_2 \in S_o)$ then multiply both sides by $y_1^2y_2^2$ obtaining $x_1y_1y_2^2 = x_2y_2y_1^2$ (since $y_1'y_1^2 = y_1$). Now $\chi(x_1)\chi(y_1)\chi(y_2)^2$ $= \chi(x_2)\chi(y_1)^2\chi(y_2)$ and dividing by $\chi(y_1)^2\chi(y_2)^2$ yields $\chi(x_1)/\chi(y_1) = \chi(x_2)/\chi(y_2)$. Thus $\chi_1 \in \hat{U}_o$. The Šilov boundary of $\ell^1(\hat{U}_o)$ is all of U_o (see Definition 2.7). We may assume that U has an identity 1 and that $1 \in U_o$ (can extend χ_1 to 1 if necessary).. Further $\ell^1(U_o)$ is a closed subalgebra of $\ell^1(U)$ and the map which restricts the maximal ideal space of $\ell^1(U)$ to $\ell^1(U_o)$ takes $\hat{U}$ onto a closed (compact) boundary of $\ell^1(U_o)$ which must contain the Šilov boundary $\hat{U}_o$. That is, there exists $\psi \in \hat{U}$ such that $\psi|U_o = \chi_1$, or $\psi|S_o = \chi$. □

6.2 Example: The hypotheses of Theorem 6.1 can not be weakened to $|\chi| = 0$ or 1. Indeed let $S = Z$, the integers under addition, let $S_o = \{0,1,2,\cdots\}$ and define $\chi \in \hat{S}_o$ by $\chi(0) = 1$, $\chi(n) = 0$ $(n > 0)$. Clearly χ is not the restriction of a character of Z. It is however the restriction of a positive-definite function on Z, namely, the function ψ, with $\psi(0) = 1$, $\psi(n) = 0$, $(n \neq 0)$.

6.3 Theorem: *Let U be an inverse semigroup and let S_o be a subsemigroup of U. Suppose $\chi \in \hat{S}_o$ then there exists a positive-definite function f on U such that $f|S_o = \chi$.*

Proof. We may assume U has an identity 1 and $1 \in S_o$ (else extend χ to be 1 at 1). The function χ is an element of $R(S_o)$ (Theorem 2.1.7) with norm 1. It corresponds to a bounded linear functional on $(\ell^1(S_o), ||\cdot||_{sp})$ and hence by the Hahn-Banach theorem extends to a bounded linear functional with norm 1 on $(\ell^1(U), ||\cdot||_{sp})$, that is, an element $\mu \in M(\hat{U})$ with $||\mu|| = 1$ (Corollary 5.2). But $1 = \chi(1) = \int_{\hat{U}} \hat{1}\, d\mu = \int_{\hat{U}} d\mu$, thus $\mu \geq 0$. Let $f(x) = \int_{\hat{U}} \hat{x}\, d\mu$, $(x \in U)$, then $f|S_o = \chi$, and f is positive-definite (Theorem 5.4). □

This theorem may be viewed as providing representing measures for multiplicative linear functionals of $\ell^1(S_o)$. Note that any positive-definite extension f of $\chi \in \hat{S}_o$ satisfies $f(x') = \overline{\chi(x)}$, $(x \in S_o)$.

§7. The regular representation

We first define the regular representation of an inverse semigroup and then restrict to subsemigroups to get any separative semigroup. We will represent $\ell^1(S)$ as an algebra of operators on $\ell^2(S)$ which algebra is isometrically isomorphic to $(\ell^1(S), ||\cdot||_{sp})$.

Recall $\ell^2(S)$ is the Hilbert space of complex functions f on S such that $||f||_2 = (\Sigma_{x\in S}|f(x)|^2)^{1/2} < \infty$. Further an orthonormal basis for $\ell^2(S)$ is given by $\{\delta_x : x \in S\}$ (see 1.1).

For $x \in U$, an inverse semigroup, e_x denotes the idempotent xx'.

7.1 Definition: For $x \in U$, an inverse semigroup, define the operator ρx on finitely supported functions on S by

$$\rho x(\delta_y) = \begin{cases} \delta_{xy} & e_x e_y = e_y \quad (\text{that is } e_x \geq e_y) \\ 0 & \text{otherwise, and} \end{cases}$$

extend to be linear.

7.2 Theorem: For $x \in U$, ρx extends to a bounded linear operator on $\ell^2(U)$ with the operator norm, $||\rho x||_{op} \leq 1$.

Proof. Choose finitely many distinct points $y_1,\cdots,y_n \in U$, numbers $c_1,\cdots,c_n \in \mathbb{C}$. Index the points so that $e_x e_{y_j} = e_{y_j}$ for $j = 1,\cdots,m$ and $e_x e_{y_j} \neq e_{y_j}$ for $j = m+1,\cdots n$ (if $m = 0$, then $\rho x \sum_{j=1}^{n} c_j \delta_{y_j} = 0$).

We claim that $j,k = 1,\cdots,m$ and $j \neq k$ implies $xy_j \neq xy_k$. Indeed $x'xy_j = e_x y_j = e_x e_{y_j} y_j = e_{y_j} y_j = y_j$, thus $xy_j = xy_k$ implies $y_j = y_k$ $(j,k = 1,\cdots,m)$. Hence

$$||\rho x \sum_{j=1}^{n} c_j \delta_{y_j}||_2 = ||\sum_{j=1}^{m} c_j \delta_{xy_j}||_2$$

$$= (\Sigma_{j=1}^{m} |c_j|^2)^{1/2} \leq (\Sigma_{j=1}^{n} |c_j|^2)^{1/2} = ||\Sigma_{j=1}^{n} c_j \delta_{y_j}||_2 .$$

Thus ρx extends to a bounded operator on $\ell^2(U)$ with $||\rho x||_{op} \leq 1$.□

7.3 <u>Corollary</u>: ρ <u>extends to be a bounded linear map of</u> $\ell^1(U)$ <u>into</u> $B(\ell^2(U))$, and $||\rho f||_{op} \leq ||f||_1$, $(f \in \ell^1(U))$.

Proof. For $f \in \ell^1(U)$ define $\rho f = \Sigma_{x \in S} f(x) \rho x$ (an absolutely convergent series in $B(\ell^2(U))$. Clearly $||\rho f||_{op} \leq ||f||_1$. □

7.4 <u>Theorem</u>: ρ <u>is a *-representation of</u> $\ell^1(U)$, <u>that is</u>: $(x,y \in U,\ f,g \in \ell^1(U))$

1) $\rho(xy) = (\rho x)(\rho y)$,

2) $\rho(f*g) = (\rho f)(\rho g)$,

3) $\rho(x') = (\rho x)*$ (* <u>denotes operator adjoint</u>)

4) $\rho(f^*) = (\rho f)*$ (see 2.7)

5) $\rho x, \rho f$ <u>are normal operators</u>

6) $||\rho f||_{op} \leq ||f||_{sp}$.

Proof. To show 1) and 2) it suffices to show $\rho x(\rho y \delta_z) = \rho(xy)\delta_z$ $(x,y,z \in U)$. Now $\rho x(\rho y \delta_z) = \delta_{xyz}$ if and only if $e_x e_{yz} = e_{yz}$ and $e_y e_z = e_z$ if and only if $e_x \geq e_z$ and $e_y \geq e_z$ if and only if $e_{xy} \geq e_z$ if and only if $\rho(xy)\delta_z = \delta_{xyz}$. Otherwise

$\rho x(\rho y\delta_z) = 0 = \rho(xy)\delta_z$.

We show 3) first for idempotents. Let $e \in E(U)$, then ρe is exactly the orthogonal projection of $\ell^2(U)$ onto $\{\Sigma_{n=1}^{\infty} c_n \delta_{y_n} : \Sigma_n |c_n|^2 < \infty,\ ey_n = y_n \text{ all } n\}$. (Another argument: by 1) we have $(\rho e)^2 = \rho e$ which together with $||\rho e||_{op} \leq 1$ implies ρe is orthogonal.) Hence $(\rho e)^* = \rho e$. For any $x \in U$ one obtains $(\rho x)(\rho x') = \rho e_x$, $(\rho x)(\rho e_x) = \rho x = (\rho e_x)(\rho x)$ and $(\rho x')(\rho e_x) = (\rho e_x)(\rho x') = \rho x'$, thus $(\rho x)^* = \rho x'$. By linear extension this proves 4). To prove 5) just observe $(\rho x)(\rho x)^* = \rho(xx') = \rho(x'x) = (\rho x)^*(\rho x)$.

Finally, let $f \in \ell^1(U)$, $n = 1,2,\cdots$ then $||(\rho f)^n||_{op}^{1/n} = ||\rho(f^n)||_{op}^{1/n} \leq ||f^n||_1^{1/n}$. Let $n \to \infty$ then the right side tends to $||f||_{sp}$. Since ρf is normal, $||(\rho f)^n||_{op}^{1/n} = ||\rho f||_{op}$ and so $||\rho f||_{op} \leq ||f||_{sp}$. □

We now set up a Plancherel theorem for $\ell^2(U)$ which will provide another way of looking at the regular representation. Recall from Theorem 2.4 that $\hat{U}$ contains an isomorphic copy of the dual (compact) group of each maximal group in U. For $x \in U$, we denote this compact group by $\Gamma(x) \subset \hat{U}$. Each such group has normalized Haar measure $m_{\Gamma(x)}$. Let m denote the direct sum of $\{m_{\Gamma(e)} : e \in E(U)\}$, thus m is carried by the union of $\{\Gamma(e) : e \in E(U)\}$ and

$$L^2(m) = \Sigma \oplus \{L^2(m_{\Gamma(e)}) : e \in E(U)\}.$$

7.5 <u>Definition</u>: For $x \in U$ define $\sigma\delta_x \in L^2(m)$ by

$$\sigma\delta_x = \begin{cases} \hat{x} & \text{on } \Gamma(x) \\ 0 & \text{otherwise,} \end{cases}$$

and extend σ to be an isometry of $\ell^2(U)$ onto $L^2(m)$ (just observe that the image of the orthonormal basis $\{\delta_x : x \in U\}$ is an orthonormal basis for $L^2(m)$; for $x \in U$, $||\sigma\delta_x||_2^2 = \int_{\Gamma(x)} |\hat{x}|^2 dm_{\Gamma(x)} = 1$ and if $y \in H(x)$, $y \neq x$, then $\int (\sigma\delta_x)\overline{(\sigma\delta_y)} dm = \int_{\Gamma(x)} \hat{x}\bar{\hat{y}} dm_{\Gamma(x)} = 0$, if $y \notin H(x)$, then $(\sigma\delta_x)\overline{(\sigma\delta_y)} = 0$.

The existence of σ is the Plancherel theorem. We now show that the regular representation of $\ell^1(U)$ is isomorphic to multiplying $L^2(m)$ by $\ell^1(U)$^.

7.6 <u>Proposition</u>: <u>Let</u> $x,y \in U$, <u>then</u> $\sigma(\rho x\delta_y) = \hat{x}\sigma\delta_y$.

Proof. If $e_x e_y = e_y$, then $\rho x\delta_y = \delta_{xy}$ and $xy \in H(y)$. Thus $\sigma(\delta_{xy}) = \hat{x}\hat{y}$ on $\Gamma(y)$ and 0 off $\Gamma(y)$, thus is identical with $\hat{x}(\sigma\delta_y)$. If $e_x e_y \neq e_y$, then $\rho x\delta_y = 0$. By the definition of $\Gamma(y)$ (see Theorem 2.4) $e_x e_y \neq e_y$ implies that $\hat{x} = 0$ on $\Gamma(y)$, and thus $\hat{x}\sigma\delta_y = 0$. □

The proposition yields the following theorem:

7.7 <u>Theorem</u>: <u>The regular representation of $\ell^1(U)$ on $\ell^2(U)$ is unitarily equivalent to the representation of $\ell^1(U)$ acting by multiplication on $L^2(m)$. The isomorphism is given by the formula</u>

$$\sigma((\rho f)g) = \hat{f}(\sigma g) \quad (f \in \ell^1(U),\ g \in \ell^2(U)).$$

It remains to show that $||\rho f||_{op} = ||f||_{sp}$, $(f \in \ell^1(U))$, but we will prove a somewhat stronger result showing that the operator norm of ρf restricted to $\ell^2(S)$, S a subsemigroup of U supporting f, is indeed equal to $||f||_{sp}$.

7.8 Definition: Let S be a subsemigroup of U, and let $f \in \ell^1(S)$. Define the operator $\rho_S f$ on $\ell^2(S)$ to be the restriction of ρf to $\ell^2(S)$. Observe that $\ell^2(S)$ is invariant under ρf, since $x,y \in S$ imply $(\rho x)\delta_y = \delta_{xy}$ or 0, which are both elements of $\ell^2(S)$. Denote the operator norm of $\rho_S f$ by $||\rho_S f||_{op}$, thus

$$||\rho_S f||_{op} = \sup\{||\Sigma_{j=1}^n c_j(\rho f\delta_{y_j})||_2 : \Sigma_{j=1}^n |c_j|^2 = 1;\ c_1,\cdots,c_n \in \mathbb{C};\ y_1,\cdots,y_n \text{ distinct points in } S,\ n = 1,2,3\cdots\}.$$

7.9 Theorem: Let S be a subsemigroup of U and let $f \in \ell^1(S)$, then $||\rho_S f||_{op} = ||f||_{sp}$.

Proof. It suffices to prove $||\rho_S f||_{op} \geq ||f||_{sp}$ for finitely supported f. Choose $x_1,\cdots,x_n \in S$ numbers $c_1,\cdots,c_n \in \mathbb{C}$ and let $f = \Sigma_{j=1}^n c_j\delta_{x_j}$. Let S_o be the subsemigroup generated by $x_1,\cdots,x_n$ and let $U_o = \bigcup\{H(y) : y \in S_o\}$. We claim U_o is a subsemigroup of U and is a union of at most 2^n-1 maximal groups in U. To see that U_o is a semigroup, let $u_1 \in H(y_1)$, $u_2 \in H(y_2)$, $y_1, y_2 \in S_o$ then $u_1u_2 \in H(y_1y_2)$ and $y_1y_2 \in S_o$. The only possible idempotents in U_o are products of $e_{x_1},\cdots,e_{x_n}$. Note that $f \in \ell^1(S_o)$ and $||\rho_{S_o} f||_{op} \leq ||\rho_S f||_{op}$, since $\ell^2(S_o) \subset \ell^2(S)$. Now $\hat{U}_o$ is a finite union of compact groups (see Theorem 2.6), and so $\hat{f}$ achieves its supremum on some $\Gamma(y)$, $y \in S_o$ (note $||f||_{sp}$ does not depend on whether f is considered as an element of $\ell^1(S_o)$,

4.7.9

$\ell^1(U_o)$ or $\ell^1(U)$). For $n = 1,2,3\cdots$

$$||(\rho_{S_o} f)^n \delta_y||_2 = ||\sigma((\rho f)^n \delta_y))||_2 = \{\int_{\Gamma(y)} |\hat{f}^n \hat{y}|^2 dm_{\Gamma(y)}\}^{1/2}$$

$$= \{\int_{\Gamma(y)} |\hat{f}|^{2n} dm_{\Gamma(y)}\}^{1/2} \leq ||\rho_{S_o} f||_{op}^n ||\delta_y||_2 \text{ (since } \delta_y \in \ell^2(S_o)).$$

Take n^{th} roots to obtain

$$\{\int_{\Gamma(y)} |\hat{f}|^{2n} dm_{\Gamma(y)}\}^{1/2n} \leq ||\rho_{S_o} f||_{op}$$

and let $n \to \infty$. The left side converges to the essential supremum of $|\hat{f}|$ relative to $m_{\Gamma(y)}$, but this number equals $||f||_{sp}$, by the choice of $\Gamma(y)$. Thus

$$||f||_{sp} \leq ||\rho_{S_o} f||_{op} \leq ||\rho_S f||_{op} \leq ||f||_{sp}. \quad \square$$

Remark: We point out that the algebraic definition of a regular representation, that is, $R(x)\delta_y = \delta_{xy}$ $(x,y \in S)$, does not in general give contractive, or even bounded, operators on $\ell^2(S)$. For example, let S be an infinite semigroup with a zero, then $R(0)(\Sigma_{j=1}^n \delta_{y_j}) = n\delta_0$ (where $n = 1,2,3\cdots$ and $y_1,\cdots,y_n$ are distinct points in S) and $||n\delta_0||_2 / ||\Sigma_{j=1}^n \delta_{y_j}||_2 = \sqrt{n}$.

A given separative semigroup S may be realized as a subsemigroup of different inverse semigroups, and the restrictions of the regular representations may be different. However, in any case the algebra of operators on $\ell^2(S)$ thus constructed is isometrically isomorphic to $(\ell^1(S), ||\cdot||_{sp})$ (Theorem 7.9). Let us call the regular representation realized on the inverse semigroup U associated to S as in Theorem 3.9 the standard regular representation of S. It can also be defined as follows:

7.10 <u>Definition</u>: Let S be a separative semigroup, then the <u>standard regular representation</u> ρ of $\ell^1(S)$ on $\ell^2(S)$ is given by:

$$\rho_x(\delta_y) = \begin{cases} \delta_{xy} & xy \in h_y \\ \\ 0 & \text{otherwise}, \ (x,y \in S). \end{cases}$$

We see that ρ is exactly the restriction of the regular representation of U to S (since $xy \in h_y$ if and only if $e_x e_y = e_y$ $(x,y \in S)$. Thus Theorems 7.7 and 7.9 show that ρ is an isometric representation of $(\ell^1(S), ||\cdot||_{sp})$ on $\ell^2(S)$.

7.11 <u>Example</u>: Let S be the subsemigroup of Z^2 (additive) generated by $x = (1,0)$ and $y = (0,1)$. As a subsemigroup of Z^2 (a group) the regular representation is $(\rho_S u)\delta_v = \delta_{u+v}$ $(u,v \in S)$. However to determine the standard regular representation note that $S = h_x \cup h_y \cup h_{x+y}$, and thus $(\rho x)\delta_y = 0$, for example, so $\rho \neq \rho_S$.

§8. Sources and related work

Hewitt and Zuckerman developed the basic ideas about the semi-simplicity of $\ell^1(S)$ and separative semigroups. Propositions 1.2, and 1.4, Section 3, and Theorem 4.3 are based on their paper [2]. In this paper also may be found a characterization of semigroups for which $\ell^1(S)$ has an identity - it is the existence of a finite set of relative units: a set of idempotents $\{e_1,\cdots,e_n\}$ such that for each $x \in S$, $e_j x = x$ for some j. For example this condition is satisfied by inverse semigroups U which are unions of finitely many groups (Theorem 2.6 shows that $\hat{U}$ is compact and so $\ell^1(U)$ has an identity).

In an earlier paper [1] Hewitt and Zuckerman investigated the algebraic structure of $\ell^1(S)$ for finite semigroups.

Theorems 2.3, 2.4 and 2.6 are closely related to some work of R. Warne and L. Williams [1] on characters of inverse semigroups. They study the lattice properties of U and $\hat{U}$ more deeply. In this connection, see also S. Šchwarz [1]. Austin [1] has further investigated duality relations between discrete inverse semigroups and compact inverse semigroups.

The Šilov boundary of $\ell^1(S)$ was investigated by W. Comfort [1] at length. The statement of Theorem 4.1 is due to him. He also showed that $\{\chi \in \hat{S}: |\chi| = 1\}$ is always contained in the Šilov boundary of $\ell^1(S)$.

In reference to Theorem 6.1 we point out that K. Ross [1] has a stronger theorem which in fact gives a necessary and sufficient condition for extension of a semicharacter: let S_o be a subsemigroup of a semigroup S, and let $\chi \in \hat{S}_o$, then $\chi \in \hat{S}|S_o$ if and only if $x \in S_o$, $y \in S$ such that $xy \in S_o$ implies that $|\chi(xy)| \leq |\chi(x)|$.

Chapter 5. Subsemigroups of locally compact abelian groups and weakly almost periodic functions

In the first section we show that the representation algebra $R(G)$ of a locally compact abelian group G is the space of Fourier-Stieltjes transforms of the measures on the dual group $\hat{G}$, Theorem 1.1. For a commutative separative semigroup S with the discrete topology, we characterize (Theorem 1.6) $\overline{R(S)}$, the closure in the sup-norm of the space $R(S)$. This result is based on being able to pair the spaces $\ell^1(S)$ and $R(S)$.

In the second section we investigate the relationship between $\overline{R(S)}$ and the space WAP(S) of weakly almost periodic functions on S. One always has $\overline{R(S)} \subset \text{WAP}(S)$. Using quasi-uniform convergence we show that $\overline{R(S)} \neq \text{WAP}(S)$ for a noncompact locally compact subsemigroup S of an LCA group, Theorem 2.12. The result for separative semigroups is given in Theorem 2.15.

We use Dunkl and Ramirez [1, Chapter 4] as a reference for needed LCA results.

§1. The representation algebra for locally compact abelian groups.

In this section, (G,+) will be a noncompact locally compact abelian group with dual group $\hat{G}$. For $\mu \in M(\hat{G})$, the Fourier-Stieltjes transform $\hat{\mu}$ of μ is defined by

$$\hat{\mu}(x) = \int_{\hat{G}} \gamma(x)\, d\mu(\gamma), \qquad (x \in G).$$

1.1 <u>Theorem</u>: $R(G) = M(\hat{G})\hat{\ }$.

Proof. Let $\mu \in M_p(\hat{G})$ so $\hat{\mu}(x) = \int_{\hat{G}}\gamma(x)\,d\mu(\gamma)$. Thus to show $(\hat{x},\mu,\hat{G}) \in S$ (where $\hat{x}(\gamma) = \gamma(x)$, $\gamma \in \hat{G}$) we need to argue that $x \mapsto \hat{x}$ is weak-* continuous from $G \to L^\infty(\mu)$.

First, for a trigonometric polynomial g on $\hat{G}$, $g = \Sigma_{i=1}^n c_i\hat{x}_i$ ($x_i \in G$, $c_i \in \mathbb{C}$), one has

$$\int_{\hat{G}} g(\gamma)\,\hat{x}(\gamma)\,d\mu(\gamma) = \Sigma_{i=1}^n c_i\hat{\mu}(x+x_i),$$

a continuous function in x. The space of trigonometric polynomials is dense in $L^1(\mu)$: to see this recall that $\hat{G}$ maps one-to-one continuously into the Bohr compactification of $\hat{G}$, the maximal ideal space of the uniform closure of the trigonometric polynomials on $\hat{G}$; thus any continuous function f on a compact subset K of $\hat{G}$ can be uniformly approximated on K by a trigonometric polynomial g without increasing the sup-norm over $\hat{G}$ of g. And so $x \mapsto \int_{\hat{G}} g\hat{x}\,d\mu$ is a continuous function for all $g \in L^1(\mu)$. Hence $\hat{\mu} \in R(G)$ and $||\hat{\mu}||_R \leq ||\mu||$. Any $\mu \in M(\hat{G})$ has the form $g\,d\nu$ ($\nu \in M_p(\hat{G})$, $g \in L^1(\nu)$).

Let $f \in R(G)$. We show there exists $\mu \in M(\hat{G})$ with $f = \hat{\mu}$ and $||\mu|| \leq ||f||_R$. Firstly, $f \in R(G_d)$ and so by Theorem 4.5.1, there exists $\mu \in M((G_d)^\wedge)$ with

$$f(x) = \int_{(G_d)^\wedge} \hat{x}(\gamma)\,d\mu(\gamma), \qquad (x \in G);$$

also μ may be chosen with $||\mu|| = ||f||_{R(G_d)}$. The space $(G_d)^\wedge$ is the Bohr compactification of $\hat{G}$. Secondly, since $f \in C^B(G) \cap R(G_d)$, Theorem 3.2.11 asserts $||f||_{R(G_d)} = ||f||_{R(G)}$. Thus we have a measure $\mu \in M((G_d)^\wedge)$ with $||\mu|| = ||f||_{R(G)}$ and $\hat{\mu}$ a continuous function on G (since it is equal to f there). It follows now from standard methods (for example, use Bochner's theorem or the proof of Eberlein's theorem, Rudin [1, p.33]) that μ is concentrated on $\hat{G}$; that is, $\mu \in M(\hat{G})$. □

1.2 Remark: There exist compact commutative semitopological semigroups S for which $WAP(S) \neq \overline{R(S)}$, (closure in sup-norm over S). This holds for $S = \overline{G}^W$, the weakly almost periodic compactification of a locally compact noncompact abelian group. There do not exist enough L^∞-representations of $\overline{G}^W$ to separate points: this follows since for locally compact noncompact abelian groups G, $WAP(G) \neq \overline{M(\hat{G})}^\wedge$ (see Dunkl and Ramirez, [1, Chapter 4]).

1.3 Corollary: Let $S = G \cup \{\infty\}$ be the one-point compactification of a locally compact noncompact abelian group G. Then $R(S) = M_o(\hat{G})^\wedge \oplus \mathbb{C}$, where $M_o(\hat{G}) = \{\mu \in M(\hat{G}) : \hat{\mu} \in C_o(G)\}$.

Proof. By Theorem 3.2.11, $M_o(\hat{G})^\wedge \oplus \mathbb{C} \subset R(S)$ (since S is of type U). The converse follows from Theorem 1.1 and the fact that each $f \in R(S)$ has a limit at ∞. □

1.4 Theorem: Let G be a locally compact abelian group and S a subsemigroup of G. For $\phi \in (S_d)^\wedge$, and $c_1,\cdots,c_n \in \mathbb{C}$, $x_1,\cdots,x_n \in S$,

$$|\Sigma_{i=1}^n c_i \phi(x_i)| \leq ||\Sigma_{i=1}^n c_i \delta_{x_i}||_{sp}$$

$$= \sup \{|\Sigma_{i=1}^n c_i \gamma(x_i)| : \gamma \in \hat{G}\} = ||\Sigma_{i=1}^n c_i \hat{x}_i||_\infty .$$

Proof. We use an argument from Arens and Singer [1]. The function ϕ defines a multiplicative linear functional on

$\ell^1(S_d) \subset \ell^1(G_d)$. But $(G_d)^\wedge$ is a boundary for $\ell^1(G_d)$, and so for $\ell^1(S_d)$ (Theorem 4.4.1). Thus

$$|\Sigma_{i=1}^n c_i\phi(x_i)| \leq \sup\ \{|\Sigma_{i=1}^n c_i\gamma(x_i)| : \gamma \in (G_d)^\wedge\}.$$

Also since $(G_d)^\wedge$ is a boundary for $\ell^1(G_d)$, given $f \in \ell^1(G_d)$, $||f||_{sp} = ||\hat{f}||_{(G_d)^\wedge} = \max\ \{|\Sigma_{x\in G} f(x)\phi(x)| : \phi \in (G_d)^\wedge\}$.

It remains to note that $\hat{G}$ is (pointwise) dense in $(G_d)^\wedge$: one way to see this is to note that $G_d \to G$ is onto and so the adjoint $\hat{G} \to (G_d)^\wedge$ has dense range. Thus for $f \in \ell^1(G_d)$, $\max\ \{|\Sigma_{x\in G} f(x)\phi(x)| : \phi \in (G_d)^\wedge\} = \max\ \{|\Sigma_{x\in G} f(x)\phi(x)| : \phi \in \hat{G}\}$.□

1.5 <u>Theorem</u>: <u>Let</u> S <u>be a subsemigroup of a locally compact abelian group</u> G <u>with dual group</u> $\hat{G}$. <u>Suppose</u> S <u>has uniformly positive</u> m_G-<u>measure</u> (<u>that is, if</u> $f \in C^B(G)$ <u>with</u> $\int_G fh\,dm_G = 0$ <u>for all</u> $h \in L^1(S) = L^1(G) \cap M(S)$, <u>then</u> $f = 0$ <u>on</u> S). <u>Then</u> $R(S) = R(G)|S$ <u>with the quotient norm</u>. <u>For example, if</u> $S = \mathbb{R}_+$, $R(\mathbb{R}_+) = R(\mathbb{R})|\mathbb{R}_+$.

Proof. Let $g \in R(G)$. Then $f = g|S \in R(S)$ and $||f||_{R(S)} \leq ||g||_{R(G)}$.

Conversely, let $f \in R(S)$ and $\varepsilon > 0$, so $f(x) = \int_\Omega (Tx)k\,d\mu = (T^*k)(x)$, $(T,\mu,\Omega) \in \mathcal{S}$ where $k \in L^1(\mu)$ with $||f||_R > ||k||_1 - \varepsilon$. We wish to show $T^*k = \hat{\lambda}|S$ for some $\lambda \in M(\hat{G})$ with $||\lambda|| \leq ||k||_1$. Extend the map $T:S \to L^\infty(\mu)$ to a contraction operator $T(||T|| \leq 1)$ on the space $M_d(S)$ of discrete measures on S. By the weak-* continuity of T and since the unit ball of $M_d(S)$ is weak-* dense in the (compact) unit ball of $M(S)$, we further extend T to a contraction operator on $M(S)$ into $L^\infty(\mu)$. Let T_1 denote the re-

5.1.5

striction of T to the subspace $L^1(S) \subset M(S)$. For $h \in L^1(S)$,

$$||T_1 h||_\infty = ||T_1 h^n||_\infty^{1/n}$$

$$\leq \liminf (||T_1|| \; ||h^n||_1)^{1/n}$$

$$\leq \liminf (||h^n||_1)^{1/n} = ||h||_{sp}$$

$$= ||\hat{h}||_\infty.$$

Hence the linear map $\rho : h \mapsto \int_\Omega (T_1 h) k d\mu$ from $L^1(S) \to \mathbb{C}$ satisfies

$$|\rho(h)| \leq ||\hat{h}||_\infty ||k||_1.$$

Since $L^1(S)^\wedge$ is a subspace of $C_o(\hat{G})$, we can extend ρ (by the Hahn-Banach theorem) to a bounded linear functional on $C_o(\hat{G})$ with no increase in norm. By the Riesz representation theorem, there exists a measure $\lambda \in M(\hat{G})$ with $||\lambda|| \leq ||k||_1 < ||f||_R + \varepsilon$, and

$$\rho(h) = \int_\Omega (T_1 h) k d\mu = \int_{\hat{G}} \hat{h} d\lambda, \quad (h \in L^1(S)).$$

Since S has uniform positive m_G-measure, it remains to show the following claim that

$$\rho(h) = \int_S (T^*k) h dm_G, \quad (h \in L^1(S)); \text{ for then}$$

$$\int_S \hat{\lambda} h dm_G = \int_G \hat{\lambda} h dm_G = \int_{\hat{G}} \hat{h} d\lambda = \rho(h) = \int_S (T^*k) h dm_G, \quad (h \in L^1(S))$$

which implies $\hat{\lambda} = T^*k = f$ on S, and $||\lambda|| < ||f||_R + \varepsilon$.

To prove the claim,, we first consider the adjoint operator $T^{**}: R(S)^* \to L^\infty(\mu)$ of the operator $T^*: L^1(\mu) \to R(S)$, (recall the remarks after Definition 2.1.2). For $x \in S$, $\delta_x \in M(S) \subset R(S)^*$, and for $h \in L^1(\mu)$,

$$\int_\Omega (T^{**}\delta_x) h d\mu = \langle h, T^{**}\delta_x \rangle = \langle T^*h, \delta_x \rangle = \int_S (T^*h) d\delta_x = (T^*h)(x)$$
$$= \int_\Omega (Tx) h d\mu. \text{ This implies that}$$

$T^{**}|L^1(S) = T_1$. Hence for $h \in L^1(S)$,

$$\begin{aligned} \rho(h) &= \int_\Omega (T_1 h) k d\mu = \int_\Omega (T^{**}h) k d\mu \\ &= \langle k, T^{**}h\rangle = \langle T^*k, h\rangle \\ &= \int_S (T^*k) h dm_G, \end{aligned}$$

the desired result. □

1.6 <u>Theorem: Let</u> S <u>be a commutative separative semigroup with the discrete topology. The space</u> $\overline{R(S)}$ <u>(closure in the sup-norm topology on</u> S) <u>consists of those bounded functions</u> $f \in \ell^\infty(S)$ <u>which satisfy the condition</u>

(*) <u>if</u> $\{\lambda_n\}_{n=1}^\infty \subset \ell^1(S)$ <u>with</u> $||\lambda_n||_1 \leq 1$ <u>and</u> $||\lambda_n||_{sp} \overset{n}{\to} 0$, <u>then</u> $\langle\lambda_n, f\rangle = \Sigma_{x\in S} f(x)\lambda_n(x) \overset{n}{\to} 0$; <u>that is, the map</u> $\lambda \mapsto \langle\lambda, f\rangle$ <u>is spectrally continuous on norm bounded subsets of</u> $\ell^1(S)$.

Proof. The proof here follows the LCA-result given in Dunkl and Ramirez [1, p. 31] and is divided into eight steps.

Let $\langle\cdot,\cdot\rangle$ denote the natural pairing between $\ell^1(S)$ and $\ell^\infty(S)$ given by $\langle\lambda, f\rangle = \Sigma_{x\in S} f(x)\lambda(x)$ $(\lambda \in \ell^1(S),\ f \in \ell^\infty(S))$.

Step 1. <u>The bilinear mapping</u> $\langle\cdot,\cdot\rangle$ <u>restricted to</u> $R(S) \subset \ell^\infty(S)$ <u>is a pairing</u>:

For $\lambda \in \ell^1(S)$, suppose $\langle\lambda, f\rangle = 0$ for all $f \in R(S)$; in particular for $f \in \hat{S}$ (2.1.7). Thus $\hat{\lambda} = 0$ (4.1.2) and so $\lambda = 0$ since $\ell^1(S)$ is semisimple (4.3.1).

Define $B_n = \{\lambda \in \ell^1(S) : ||\lambda||_1 \leq n\}$, $n \in Z_+$. The w-topology on $\ell^1(S)$ is defined by $\lambda_\alpha \overset{\alpha}{\to} 0$ in w if and only if $\langle\lambda_\alpha, f\rangle \overset{\alpha}{\to} 0$ for all $f \in R(S)$. Let U be the inverse semigroup with $S \subset U$ as in Chapter 4. We assume U has an identity and so $\hat{U}$ is compact.

5.1.6

Step 2. <u>For</u> $\lambda \in \ell^1(S)$, $||\lambda||_1 = \sup \{|<\lambda,f>| : ||f||_\infty \leq 1,\ f \in \mathcal{R}(S)\}$:

Choose $\varepsilon > 0$ and let F be a finite (hence compact) subset of S which supports most of λ. Since $\overline{\mathcal{R}(S)}$ is a commutative C*-algebra $\overline{\mathcal{R}(S)} \cong C(\kappa S)$ where κS is a compact space with S dense in it. Find $g \in \overline{\mathcal{R}(S)}$ with $||g||_\infty \leq 1$ and $|<\lambda,g>| > ||\lambda||_1 - \varepsilon$. Choose $f \in \mathcal{R}(S)$ with $\sup \{|(f-g)(x)| : x \in F\} < \varepsilon$ and $||f||_\infty \leq 1$. Thus

$$|<\lambda,f>| \geq |<\lambda,g>| - |<\lambda,f-g>|$$

$$\geq ||\lambda||_1 - \varepsilon - ||\lambda||_1\ \varepsilon,$$

which is arbitrarily close to $||\lambda||_1$.

Step 3. B_n <u>is w-closed</u>:

Note the inequality $|<\lambda,f>| \leq ||\lambda||_1\ ||f||_\infty$ $(\lambda \in \ell^1(S),$ $f \in \mathcal{R}(S))$. Secondly we observe that if $\{\lambda_\alpha\} \subset B_n$ and $\lambda_\alpha \overset{\alpha}{\to} \lambda$ in $w(\lambda \in \ell^1(S))$, then $|<\lambda,f>| = \lim_\alpha |<\lambda_\alpha,f>| \leq ||\lambda_\alpha||_1 ||f||_\infty$ $\leq n||f||_\infty$, $(f \in \mathcal{R}(S))$, and so $||\lambda||_1 \leq n$ by Step 2.

Step 4. B_n <u>is w-bounded</u>:

Let $f \in \mathcal{R}(S)$ and let $V = \{\lambda \in \ell^1(S) : |<\lambda,f>| \leq 1\}$. The set V is a typical subbasic w-neighborhood of 0. Let $\rho = n||f||_\infty$, then $B_n \subset \rho V$.

Let $\mathcal{T}$ be the topology on $\mathcal{R}(S)$ of uniform convergence on the norm balls B_n of $\ell^1(S)$.

Step 5. $\mathcal{T}$ <u>is equivalent to the sup-norm topology on</u> S:

Note $||f||_\infty = \sup \{|<\lambda,f>| : \lambda \in B_1\}$, $(f \in \mathcal{R}(S))$.

5.1.6

Step 6. <u>For</u> $f \in \ell^\infty(S)$, $f \in \overline{R(S)}$ <u>if and only if the linear functional</u> $\lambda \mapsto \langle\lambda, f\rangle$ <u>is</u> w-<u>continuous on</u> B_n:

The norm balls B_n are convex, circled, w-closed, w-bounded, and $\bigcup_{n=1}^{\infty} B_n = \ell^1(S)$. We apply the Grothendieck completion theorem (Köthe [1, p. 269]) to characterize the T (sup-norm on S) completion of $R(S)$ (that is, $\overline{R(S)}$) as the linear functionals on $\ell^1(S)$ which are w-continuous on each B_n.

Embed $\ell^1(S)$ into $C(\hat{U})$ by taking the Fourier transforms.

Step 7. <u>The weak</u> $\sigma(C(\hat{U}), M(\hat{U}))$-<u>topology on</u> $\ell^1(S)$ <u>is equivalent to the</u> w-<u>topology</u>:

Let $\{\lambda_\alpha\} \subset \ell^1(S)$. Suppose $\lambda_\alpha \overset{\alpha}{\to} 0$ in w. Now for $\mu \in M(\hat{U})$,

$$\int_{\hat{U}} \hat{\lambda}_\alpha d\mu = \int_{\hat{U}} (\Sigma_{x \in S} \chi(x) \lambda_\alpha(x)) d\mu(\chi)$$

$$= \Sigma_{x \in S} (\int_{\hat{U}} \chi(x) d\mu(\chi)) \lambda_\alpha(x)$$

$$= \Sigma_{x \in S} \hat{\mu}(x) \lambda_\alpha(x) = \langle\lambda_\alpha, \hat{\mu}\rangle \overset{\alpha}{\to} 0,$$

since $\hat{\mu} \in M(\hat{U})^\wedge|S = R(U)|S = R(S)$ (Corollary 4.5.2 and Theorem 4.5.3). Thus $\lambda_\alpha \overset{\alpha}{\to} 0$ weakly in $C(\hat{U})$.

Conversely, suppose $\lambda_\alpha \overset{\alpha}{\to} 0$ weakly in $C(\hat{U})$. Let $f \in R(S)$. Choose $\mu \in M(\hat{U})$ with $\hat{\mu}|S = f$, so

$$\langle\lambda_\alpha, f\rangle = \langle\lambda_\alpha, \hat{\mu}\rangle = \Sigma_S \hat{\mu}\lambda_\alpha$$

$$= \int_{\hat{U}} \hat{\lambda}_\alpha d\mu \overset{\alpha}{\to} 0.$$

Thus the topologies are equivalent on B_n.

Step 8. <u>The linear functional</u> $\Phi: \lambda \mapsto \langle\lambda, f\rangle$ <u>is</u> w-<u>continuous on</u> B_n <u>if and only if</u> Φ <u>is spectrally continuous on</u> B_n:

5.1.7

Let K be the kernel of the linear functional Φ. The functional Φ is w-continuous on B_n if and only if Φ is weakly continuous on B_n ($\subset C(\hat{U})$) if and only if $K \cap B_n$ is weakly closed in B_n if and only if $K \cap B_n$ is closed in B_n (weak and strong closures are the same on convex sets) if and only if Φ is sup-norm-on-$\hat{U}$ continuous on B_n if and only if Φ is spectrally continuous on B_n, ($||\hat{f}||_\infty = ||f||_{sp}$, $f \in \ell^1(S)$ from Theorem 4.4.1).

To finish the proof simply combine Steps 6 and 8. □

1.7 <u>Example</u>: Recall the semigroup S constructed in 3.1.11. Let G_0 and S be compact. There is a canonical homomorphism $\pi : S \to G_0$ given by $\pi x = \pi_{\alpha 0} x$ (for $x \in G_\alpha$, $\alpha \in \Lambda$). Thus there is an induced map $\pi^* : C(G_0) \to C(S)$, and for any $f \in C(S)$, $f - \pi^*(f|G_0) \in C(S)$ and is 0 on G_0. Thus $f - \pi^*(f|G_0)$ vanishes at ∞ on each G_α, $\alpha \neq 0$, and we can decompose $C(S) = \pi^* C(G_0) \oplus \Sigma\oplus_{\alpha > 0} C_o(G_\alpha)$.

For an LCA group G, let $R_o(G) = R(G) \cap C_o(G) = M_o(\hat{G})^\wedge$ (see 1.3). We claim $R(S) = \pi^* R(G_0) \oplus \Sigma\oplus_{\alpha > 0} R_o(G_\alpha)$ (splitting in $C(S)$), thus $R(S)$ separates the points of S.

Let $f \in R(S)$, then $f|G_0 \in R(G_0)$ and $\pi^*(f|G_0) \in R(S)$ (since π is a homomorphism $S \to G_0$). Thus $f - \pi^*(f|G_0) \in R(S)$ and vanishes at ∞ on each G_α, $\alpha > 0$. For $\alpha > 0$, let $f_\alpha = (f - \pi^*(f|G_0))|G_\alpha$, then $f_\alpha \in R(G_\alpha) \cap C_o(G_\alpha) = R_o(G_\alpha)$ (note G_α is a subsemigroup of S). Thus $R(S)$ is contained in the direct sum.

For the converse, we already know $\pi^* R(G_0) \subset R(S)$, so it suffices to show $R_o(G_\alpha) \subset R(S)$ (interpreting $f \in R_o(G_\alpha)$ to be 0 off G_α), $\alpha > 0$.

Fix $\alpha > 0$, and $f \in R_o(G_\alpha)$ and consider $E = \bigcup_{\beta \geq \alpha} \hat{G}_\beta$ as a subsemigroup of $\hat{S}_d$ (see 4.2.4). There exists a unique measure $\mu_\alpha \in M_o(\hat{G}_\alpha)$ so that $\hat{\mu}_\alpha(x) = f(x)$, $(x \in G_\alpha)$. By induction on the number of γ such that $\beta > \gamma \geq \alpha$, define $\mu_\beta \in M(\hat{G}_\beta)$ by

$$\mu_\beta = -\Sigma_{\alpha \leq \gamma < \beta}(\hat{\pi}_{\beta\gamma})^* \mu_\gamma$$

(note $(\hat{\pi}_{\beta\gamma})^*$ is the induced homomorphism $M(\hat{G}_\gamma) \to M(\hat{G}_\beta)$). Now define $\mu \in M(E) \subset M(\hat{S}_d)$ to be μ_β on $\hat{G}_\beta$ $(\beta \geq \alpha)$. It is easy to check that $\int_E \hat{x} d\mu$ is 0 off G_α and equals $f(x)$ for $x \in G_\alpha$. By Corollaries 4.5.2 and 3.2.12, $f \in R(S)$.

§2. Weakly almost periodic functions

In this section, we investigate the relationship between $R(S)$ and $WAP(S)$ on the commutative semitopological semigroup S.

2.1 Definition: A bounded continuous function f on S is said to be weakly almost periodic if and only if the orbit $O(f)$ of $f = \{f_y : y \in S\}$ is relatively weakly compact in the space $C^B(S)$ of continuous bounded functions on S. The set of all such f is denoted by $WAP(S)$.

2.2 Remark: The set $WAP(S)$ is a translation invariant subspace of $C^B(S)$ containing $R(S)$ (2.1.4).

2.3 Theorem (Eberlein [1]): $WAP(S)$ is a closed subspace of $C^B(S)$.

Proof. Let $\{f_n\}_{n=1}^\infty \subset WAP(S)$ be a sequence from $WAP(S)$ such that $f_n \overset{n}{\to} f$ uniformly in $C^B(S)$, $(f \in C^B(S))$. To show $f \in WAP(S)$ it suffices (by the Eberlein-Šmulian theorem, Dunford and Schwartz

[1, p. 430]) to show O(f) is relatively weakly countably compact. Thus let $\{s_i\}$ be a sequence in S. We extract from $\{s_i\}$ via a diagonal process a subsequence $\{s_j\}$ such that $(f_n)_{s_j}$ has a weak limit g_n for each n. Now

$$\begin{aligned} ||g_n - g_m||_\infty &= \sup \{|\int_S (g_n - g_m)\, d\mu| : \mu \in M(S),\ ||\mu|| \le 1\} \\ &= \sup \{|\lim_j \int_S ((f_n)_{s_j} - (f_m)_{s_j})\, d\mu| : \mu \in M(S),\ ||\mu|| \le 1\} \\ &\le ||f_n - f_m||_\infty . \end{aligned}$$

Thus $\{g_n\}$ is a Cauchy sequence in $C^B(S)$ with a strong (and hence weak) limit; call it g. It now follows that g is a weak cluster point of $\{f_{s_j}\}$. □

2.4 <u>Theorem</u> (Eberlein [1]): <u>WAP(S) is a closed subalgebra of</u> $C^B(S)$.

Proof. Let $f, g \in WAP(S)$. Pick $\{s_n\} \subset S$. Choose a subsequence $\{s_i\} \subset \{s_n\}$ and $f', g' \in C^B(S)$ with $f_{s_i} \xrightarrow{i} f'$ and $g_{s_i} \xrightarrow{i} g'$ weakly. View $C^B(S)$ as the space of continuous functions on the Stone-Čech compactification βS of S. Thus $(fg)_{s_i} \xrightarrow{i} f'g'$ pointwise on βS, and hence weakly in $C^B(S) = C(\beta S)$ by the Lebesgue dominated convergence theorem. □

We now introduce the useful concept of quasi-uniform convergence in $C^B(S)$.

2.5 <u>Definition</u>: A net of functions $\{f_\alpha\}_{\alpha \in A}$ in $C^B(S)$ is said to converge <u>quasi-uniformly on S</u> to $f \in C^B(S)$ if and only if $f_\alpha \xrightarrow{\alpha} f$ pointwise on S and for all $\varepsilon > 0$ and $\alpha_0 \in A$, there exist $\alpha_1, \cdots, \alpha_k \ge \alpha_0$ such that for each $x \in S$,

5.2.7

$$\min \{|f_{\alpha_i}(x)-f(x)|:1 \le i \le k\} < \varepsilon.$$

2.6 Proposition: A bounded sequence $\{f_n\}$ from $C^B(S)$ converges weakly to $f \in C^B(S)$ if and only if $\{f_n\}$ and every subsequence of $\{f_n\}$ converge to f quasi-uniformly on S.

Proof. See Dunford and Schwartz [1, p. 281]. □

2.7 Example: Let $S = (Z,\min)$. Then $WAP(S) = \overline{R(S)}$ (closure in $C^B(S)) = C(\{-\infty\} \cup Z \cup \{+\infty\})$.

Proof. That $WAP(S) \supset \overline{R(S)}$ follows from 2.1.4 and Theorem 2.3. That $\overline{R(S)} \supset C(\{-\infty\} \cup Z \cup \{+\infty\})$ follows since the characteristic functions of the sets $\{-\infty\} \cup \{n \in Z: n \le k\}$ are all in $R(S)$, $(k \in S)$.

Let $f \in WAP(S)$. Suppose $\lim_{n\to+\infty} f(n)$ does not exist. Then there are sequences $\{n_i\}$, $\{m_j\}$ with $n_i \xrightarrow{i} +\infty$, $m_j \xrightarrow{j} +\infty$ and $f(n_i) \xrightarrow{i} a$, $f(m_j) \xrightarrow{j} b$, and $a \ne b$. Since f is weakly almost periodic there exists a function $g \in C^B(S)$ with $f_{n_i} \xrightarrow{i} g$ weakly (passing to a subsequence if necessary). But $f_{n_i} \xrightarrow{i} f$ pointwise (remember $nm = m$ for $n \ge m$, $(n,m \in S)$); and so $f_{n_i} \xrightarrow{i} f$ weakly. Let $\varepsilon = |a-b|$ and let N,M be chosen such that $|f(n_i)-a| < \varepsilon/3$ and $|f(m_j)-b| < \varepsilon/3$ for $n_i \ge N$, $m_j \ge M$. By Proposition 2.6, there exist $n_1,n_2,\cdots,n_k \ge N$ with

$$\min \{|f_{n_i}(\ell)-f(\ell)|:1 \le i \le k\} < \varepsilon/3, \ (\ell \in S).$$

Let m_j be chosen $\ge M$ and each n_i $(1 \le i \le k)$, then

$$|f_{n_i}(m_j)-f(m_j)| = |f(n_i)-f(m_j)| \ge \varepsilon/3,$$

a contradiction. Thus $\lim_{n\to+\infty} f(n)$ exists.

Suppose $\lim_{n\to-\infty} f(n)$ does not exist. Then there exist sequences $\{n_i\}$, $\{m_j\}$ with $n_i \xrightarrow{i} -\infty$, $m_j \xrightarrow{j} -\infty$ and $f(n_i) \xrightarrow{i} a$, $f(m_j) \xrightarrow{j} b$, and $a \neq b$. Let $\varepsilon = |a-b|$ and let N,M be chosen such that $|f(n_i)-a| < \varepsilon/3$ and $|f(m_j)-b| < \varepsilon/3$ for $n_i \geq N$, $m_j \geq M$. Since f is weakly almost periodic, there exists a function $g \in C^B(S)$ with $f_{n_i} \xrightarrow{i} g$ weakly (passing to a subsequence if necessary). But $f_{n_i}(\ell) = f(n_i)$ for $n_i \leq \ell$, and so $f_{n_i} \xrightarrow{i} a$ pointwise on S; thus $g = a$. By Proposition 2.6, there exist $n_1,\cdots,n_k \geq N$ with

$$\min \{|f_{n_i}(\ell)-a| : 1 \leq i \leq k\} < \varepsilon/3, \quad (\ell \in S).$$

Let m_j be chosen with $m_j < \min \{n_i : 1 \leq i \leq k\}$, then $\min \{|f_{n_i}(m_j)-a| : 1 \leq i \leq k\} = \min \{|f(m_j)-a| : 1 \leq i \leq k\} \geq \varepsilon/3$, a contradiction. Thus $f \in C(\{-\infty\} \cup Z \cup \{+\infty\})$. □

2.8 <u>Example</u>: a) Let $S = (Z_+, \min)$. Then $WAP(S) = \overline{R(S)} = C(Z_+ \cup \{\infty\})$.

b) Let $S = \{Z_+, \max\}$. Then $WAP(S) = \overline{R(S)} = C(Z_+ \cup \{\infty\})$.

c) If S is compact, then $WAP(S) = C(S)$, (Burckel [1, p. 2]).

2.9 <u>Proposition</u>: *Let S be a locally compact subsemigroup of a locally compact abelian group G. If S is not a group, then there exists a closed subgroup $H \subset G$ with H isomorphic to the integers Z and an $x \in H$ with $x \in S \cap H$ and $-x \notin S$, indeed S contains a copy of $Z_+ \setminus \{0\}$.*

Proof. Firstly, note that if T is a dense locally compact subsemigroup of a locally compact abelian group K, then $T = K$:

T a dense locally compact subspace of a Hausdorff space is therefore open, and so if there exists an $x \in K$ with $x \notin T$, then $x-T$ is open in $K \setminus T$, a contradiction.

Secondly, note that a locally compact subsemigroup T of a compact group is a compact group: $\overline{T}$ is a compact subsemigroup of K and so $\overline{T}$ is a cancellative semigroup; and so by 1.1.8, $\overline{T}$ is a group. Now use the first part to show $T = \overline{T}$.

Finally, pick $x \in S$ and $-x \notin S$. Let H be the closed group generated by x; that is, $H = \overline{Gp}\ \{x\} \subset G$. Since H is monothetic either $H = Z$ or H is compact, (Rudin [1, p. 39]). If H is compact, then $S \cap H$ is a locally compact subsemigroup of the compact group H, and so $S \cap H$ is also a compact group, a contradiction. Thus $H = Z$. □

2.10 Remark: For G a noncompact locally compact abelian group, $WAP(G) \neq \overline{R(G)} = \overline{M(\hat{G})^{\wedge}}$, (see Dunkl and Ramirez [1, p. 45]).

We extend this result to a class of discrete semigroups. We will adapt the proof that $WAP(Z) \neq \overline{R(Z)}$ from Rudin [4].

2.11 Definition: Let S be a discrete semigroup and let $E \subset S$ such that $(E+x_1) \cap (E+x_2)$ is finite for all $x_1, x_2 \in S$ with $x_1 \neq x_2$. We call E a T-set.

2.12 Theorem: Let G be an LCA group and S a noncompact locally compact subsemigroup of G. Then $WAP(S) \neq \overline{R(S)}$.

Proof. The proof will be divided into five steps. By Remark 2.10 we may assume S is not a group, and so by Proposition 2.9, G contains a copy of Z with $Z_+ \setminus \{0\} \subset S$ and $Z_- \setminus \{0\}$ disjoint from S.

5.2.12

Let $E = \{nk!:1 \leq n \leq k,\ k = 1,2,\cdots\}$ a T-set in Z (Dunkl and Ramirez [1, p. 42]). Let V be a relatively compact symmetric neighborhood of $0 \in G$ with $(V+V) \cap Z = \{0\}$; and let W be a symmetric neighborhood of 0 with $W \subset W+W \subset V$.

Let c be a bounded function on E and $u \in C^B(G)$ with $0 \leq u \leq 1$, spt $u \subset W$, and $u(0) = 1$. Let $f \in C^B(G)$ be defined by

$$f(x) = \begin{cases} c(y)u(x-y), & x \in y+W,\ y \in E \\ 0 & , \text{otherwise} \end{cases} .$$

We begin to show $f \in WAP(G)$. Let $\{s_n\}_{n=1}^{\infty}$ be a sequence from G. If $\{s_n\}$ has a cluster point $s \in G$, then $\{f_{s_n}\}$ has as a weak (indeed, uniform) cluster point the function f_s, (since u and hence f are uniformly continuous). Thus we may assume for all i,j that $s_i - s_j \notin V$ for $i \neq j$ (by passing to a subsequence). Now spt $f \subset E+W$.

Step 1. <u>If</u> $i \neq j$, <u>then</u> spt $f_{s_i} \cap$ spt f_{s_j} <u>is compact</u>:

Let $x \in (\text{spt } f - s_i) \cap (\text{spt } f - s_j)$. Write $x = a+w-s_i = b+w'-s_j$ $(a,b \in E,\ w,w' \in W)$, So $a-b = w'-w+s_i-s_j \in V + (s_i-s_j)$. Now $a-b \in E - E \subset Z$ and $a-b \neq 0$ (since $s_i - s_j \notin V = -V$). Denote the unique integer in $V + (s_i-s_j)$ by t, (t is unique since $(V+V) \cap Z = \{0\}$). Also $t \neq 0$. Write $x = a+w-s_i = b+t+w-s_i \in E + t+W-s_i$. But $x = a+w-s_i \in E + W-s_i$. So

$$x \in (E \cap (E+t)) + W-s_i,$$

a relatively compact set (a union of finitely many translates of W).

5.2.12

Step 2. _If_ $f(x+s_i) \neq 0 \neq f(y+s_i)$ _for infinitely many_ i, _then_ $x-y \in V$:

For $f(x+s_i) \neq 0$, we may write $x+s_i = a_i+w_i$ $(a_i \in E,\ w_i \in W)$. Similarly, write $y+s_i = b_i + w_i'$ $(b_i \in E,\ w_i' \in W)$. Now

$$a_i-b_i = x-y-w_i + w_i' \in (x-y) + V.$$

Let t be the unique integer in $(x-y) + V$. So for s_i, s_j $(i \neq j)$ satisfying the condition,

$$x+s_i = a_i+w_i, \quad \text{and}$$
$$x+s_j = a_j+w_j;$$

thus

$$a_i-a_j \in (s_i-s_j) + V,$$

which implies $a_i \neq a_j$. Also

$$a_i = b_i+t \in E+t,$$

so

$$\{a_i\} \subset E \cap (E+t),$$

which means $t = 0$ since E is a T-set in Z. Hence $0 \in (x-y) + V$, or $x-y \in -V = V$.

Step 3. _Suppose there exists a subsequence_ $\{s_j\}$ _of_ $\{s_i\}$ _with_ $f_{s_j} \overset{j}{\to} 0$ _pointwise on_ G. _Then_ $f_{s_j} \overset{j}{\to} 0$ _weakly in_ $C^B(G)$:

Let $N \in Z_+$ and $\varepsilon > 0$. Let $n_1 = N$ and $n_2 = N+1$. By Step 1,

$$\min \{|f_{s_\ell}(x)| : \ell = n_1, n_2\} = 0$$

off a compact set $K \subset G$. For each $k \in K$, there exists a $j \geq N$ with $|f_{s_j}(k)| < \varepsilon$ and so there exists a neighborhood V_k of k with

5.2.12

$$\sup\{|f_{s_j}(y)| : y \in V_k\} < \varepsilon.$$

By compactness of K, there exists $n_3, \cdots, n_m \geq N$ with

$$\min\{|f_{s_\ell}(y)| : \ell = n_1, \cdots, n_m\} < \varepsilon, \quad (y \in G).$$

Similarly for any subsequence of $\{s_j\}$, and so by Proposition 2.6, $f_{s_j} \overset{j}{\to} 0$ weakly.

Step 4. $f \in WAP(G)$:

It remains to consider the case that there exists $x_0 \in G$ with $\limsup_i |f_{s_i}(x_0)| \neq 0$. By passing to a subsequence if necessary, we assume $f_{s_i}(x_0) \neq 0$ (all i). By Step 2 if $y \in G$ is such that $\limsup_i |f_{s_i}(y)| \neq 0$, then $y \in x_0 + V$, a relatively compact subset of G. Let U be a relatively compact open neighborhood of $x_0 + \bar{V}$. Since f is uniformly continuous, $\{f_{s_i}\}$ is an equicontinuous family on $\bar{U}$. Hence there exists a subsequence $\{s_j\} \subset \{s_i\}$ with $\{f_{s_j}\}$ converging uniformly (hence weakly) to $g \in C(\bar{U})$. Note $g = 0$ on $\bar{U} \setminus (x_0+\bar{V})$. By Step 3, $f_{s_j} \overset{j}{\to} 0$ weakly on $G \setminus U$; hence $\{f_{s_j}\}$ converges weakly to the function which is g on U and 0 off U, an element of $C^B(G)$.

Clearly $f|S \in WAP(G)|S \subset WAP(S)$. We will need the deep Hardy and Littlewood inequality (Zygmund [1, p. 199]) that

$$|\Sigma_{n=1}^{k} e^{in \log n} e^{in x}| \leq Ck^{1/2},$$

$(0 \leq x \leq 2\pi, k = 1,2,\cdots)$. We let the bounded function c on E be defined by

5.2.14

$$c(y) = e^{-in \log n},\ y = nk! \in E(1 \le n \le k,\ k = 1,2,\cdots).$$

Step 5. <u>With this choice of</u> c, $f \notin \overline{R(S)} \supset \overline{M(\hat{G})^{\wedge}}|S$:

We show that $f|Z_+ \notin \overline{R(Z_+)} \supset \overline{R(S)}|Z_+$. Let $\mu_k \in \ell^1(Z_+)$ be defined by

$$\mu_k = \frac{1}{k} \Sigma_{n=1}^{k} e^{in \log n} \delta_{nk!} \quad ,$$

$(k = 1,2,\cdots)$. Then $||\mu_k||_1 = 1$ and $||\mu_k||_{sp} \le Ck^{-1/2} \overset{k}{\to} 0$ (Theorem 1.4). But $\langle \mu_k, f|Z_+\rangle = 1 \overset{k}{\not\to} 0$. Hence $f|Z_+ \notin \overline{R(Z_+)}$ (by Theorem 1.6). □

2.13 <u>Corollary</u>: <u>Let</u> G <u>be an</u> LCA <u>group and</u> S <u>a noncompact locally compact subsemigroup of</u> G. <u>Then there exists</u> $f \in WAP(G)$ <u>with</u> $f|S \notin \overline{R(S)}$.

Proof. All we have to note is that if S is a group and $f \in WAP(S)$, then f extends to a function in WAP(G), (deLeeuw and Glicksberg [2], Burckel [1, p. 49]). □

In the final two results the semigroups are discrete.

2.14 <u>Theorem</u>: <u>Let</u> H(x) <u>be a maximal subgroup of the inverse semigroup</u> U. <u>Let</u> $f \in WAP(H(x))$. <u>Define</u> $g^b \in \ell^\infty(U)$ <u>by</u>

$$g^b(y) = \begin{cases} g(e_x y), & e_y \ge e_x \\ 0, & e_y e_x = e_x \end{cases},$$

<u>for</u> $g \in \ell^\infty(H(x))$. <u>Then</u> $f^b \in WAP(U)$.

Proof. Let $\{y_n\}$ be a sequence from U. Suppose for infinitely many n that $e_{y_n} e_x \neq e_x$. Then $f^b_{y_n} = 0$ for these n, so 0 is a weak cluster point of $\{f^b_{y_n}\}$.

We may thus assume $e_{y_n} e_x = e_x$ for all n. Now $\{f_{e_x y_n}\}$ has a weak cluster point g_o in $\ell^\infty(H(x))$. We now note that $\{f^b_{y_n}\}$ has as a weak cluster point in $\ell^\infty(U)$ the function equal to g^b_o, since $g \mapsto g^b : \ell^\infty(H(x)) \to \ell^\infty(U)$ is a (weakly) continuous linear map. Thus $(f^b)_{y_n} = (f_{e_x y_n})^b$ has $(g_o)^b$ as a weak cluster point. Hence $f \in WAP(U)$. □

2.15 Theorem: Let S be a separative semigroup with at least one h_x infinite. Then $WAP(S) \neq \overline{R(S)}$.

Proof. Let S be a subsemigroup of the associated (4.3.9) inverse semigroup U, and so h_x is a subsemigroup of the group $H(x)$. By Corollary 2.13, there exists $f \in WAP(H(x))$ with $f \notin \overline{R(h_x)}$. Extend f by Theorem 2.14 to $f^b \in WAP(U)$. Thus $f^b|S \in WAP(U)|S \subset WAP(S)$. Also

$$f \notin \overline{R(h_x)} = \overline{R(S)|h_x} \supset \overline{R(S)}|h_x \quad \text{(Theorem 4.5.3)}.$$

Thus $f^b|S \notin \overline{R(S)}$. □

2.16 Remark: Theorems 2.12 and 2.15 yield a large class of semigroups S for which the weak almost periodic compactification $\bar{S}^W$ of S is different from the $R(S)$ compactification $\bar{S}^R$ of S; (the space $\bar{S}^R$ is the closure of the image of S in $A(S)$, the dual of $R(S)$ under ρ, see 2.1.11, and is also the maximal ideal space of $\overline{R(S)}$).

Chapter 6. Representations in Q-algebras

We will put forth a somewhat novel way of investigating commutative semigroups. In the past, much emphasis was placed on the theory of semicharacters. This theory has at least two inadequacies. There are not enough semicharacters to separate points in the case of a compact semitopological, but not topological, semigroup and in the case of a nonseparative semigroup (considered discrete). Our theory of L^∞-representations goes some way toward alleviating the first difficulty, but is helpless in the face of the second.

The axe we will grind in this chapter is the idea of homomorphisms of Z_+^n into a semigroup. Granted, everybody knows that any commutative semigroup is a quotient of some power of Z_+, but it seems no one has tried to use harmonic analysis on Z_+^n to investigate general semigroups. The idea is to extend a homomorphism of Z_+^n to a homomorphism of P_n, complex polynomials in n variables, onto the semigroup algebra of S. We will then look at those functions on S which induce linear functionals on P_n which are bounded in the sup-norm of P_n over the product of unit discs, that is, the semigroup of characters of Z_+^n. This concept is well suited for studying quotient semigroups, like the nil-thread. The upshot of this approach is that one is led into representing a semigroup in the unit ball of the quotient of a function algebra modulo a closed ideal (roughly, the nearest thing to an L^∞-space that can have a nontrivial radical). Further one obtains an algebra of functions having many properties of the classical

Fourier-Stieltjes transforms on groups.

At first glance the reader may think that by looking at homomorphisms of Z_+^n into the semigroup one loses the topology, but we preserve it by requiring the above-mentioned functions to be continuous.

In the first section we collect information about quotient algebras and weak topologies on them. In the second section we present Q-representations and the algebra $RQ(\dot{S})$, as discussed above. Relations between $RQ(S)$ and $R(S)$ will be taken up in the subsequent chapter. In the important case of a semigroup with a dense inverse semigroup RQ and R are identical. Naturally we expect important differences between RQ and R on something like the nil-thread, and for this also see the next chapter.

By the way, as a consequence of the P_n-homomorphism idea we are able to give a quick proof of Redéi's theorem on finitely generated commutative semigroups (see Chapter 7).

§1. Q- algebras

Recall from Definition 1.2.15 that a function algebra A is a uniformly closed subalgebra of $C_o(X)$ which separates the points of X (where X is a locally compact Hausdorff space). The space of multiplicative linear functionals is denoted by M_A.

1.1 <u>Definition</u>: Let θ be an index set, then Z_+^θ denotes the additive semigroup of Z_+-valued functions on θ with only finitely many positive values, and the remaining values zero. Let S be a (multiplicative, commutative) semigroup with unit 1, and let $\{x_j : j \in \theta\} \subset S$, then for $\alpha \in Z_+^\theta$, define

6.1.5

$$x^{\alpha} = \Pi_{j \in \theta} x_j^{\alpha_j}, \quad (x^0 = 1).$$

1.2 Remark: In the above definition, the map $\alpha \mapsto x^{\alpha}$ is a homomorphism of Z_+^{θ} into S, and there is a one-to-one correspondence between such homomorphisms and θ-indexed subsets of S.

We recall that $\mathbb{T}$ denotes the unit circle $\{\lambda \in \mathbb{C}: |\lambda| = 1\}$, U denotes the open unit disc $\{\lambda \in \mathbb{C}: |\lambda| < 1\}$, and $\overline{U}$ denotes the closed unit disc.

1.3 Definition: Let θ be an index set, then $\mathbb{T}^{\theta}$, respectively $\overline{U}^{\theta}$, denotes the space of $\mathbb{T}$, respectively $\overline{U}$, -valued functions on θ, furnished with the Tikhonov (pointwise) topology. Then $\mathbb{T}^{\theta}$ is a compact topological group and $\overline{U}^{\theta}$ is a compact topological semigroup (under the pointwise operations).

1.4 Notation: For $j \in \theta$, we will use z_j to denote the $j^{\underline{th}}$ coordinate function on $\mathbb{T}^{\theta}$ and $\overline{U}^{\theta}$. For $\alpha \in Z_+^{\theta}$, z^{α} is called a monomial (note z^{α} is a continuous function on $\mathbb{T}^{\theta}$ and $\overline{U}^{\theta}$).

1.5 Definition: Let θ be an index set, then P_{θ} denotes the linear span of $\{z^{\alpha}: \alpha \in Z_+^{\theta}\}$, that is, the polynomial functions, and A_{θ} denotes the uniform closure of P_{θ} in $C(\mathbb{T}^{\theta})$. Elements of P_{θ} can be written in the form $\Sigma_{\alpha \in Z_+^{\theta}} a_{\alpha} z^{\alpha}$, where $a_{\alpha} \in \mathbb{C}$ and only finitely many a_{α}'s are nonzero. Also A_{θ} is a function algebra on $\mathbb{T}^{\theta}$ and is called the θ-polydisc algebra. Note that $1 = z^0 \in A_{\theta}$. The sup-norm in A_{θ} is denoted $||\cdot||_{\infty}$.

1.6 Proposition: The maximal ideal space of A_θ may be identified with $\bar{U}^\theta$, and the Šilov boundary is $\mathbb{T}^\theta$.

Proof. Let $\psi \in M_{A_\theta}$, then the map $j \mapsto \psi(z_j)$, $(j \in \theta)$ defines a point in $\bar{U}^\theta$, since $||z_j||_\infty = 1$. Conversely, let $p = \Sigma_\alpha a_\alpha z^\alpha$ be a polynomial and let $\lambda \in \bar{U}^\theta$, then $|p(\lambda)| = |\Sigma_\alpha a_\alpha \lambda^\alpha| \leq ||p||_\infty$ by the maximum modulus principle for analytic functions of finitely many complex variables (Rudin [3, p. 4]). Thus each element of A_θ may be considered as a continuous function on $\bar{U}^\theta$. To see that $\mathbb{T}^\theta$ is the Silov boundary, let $\lambda \in \mathbb{T}^\theta$, and let E be a finite subset of θ. Define $p_E = \Sigma_{j \in E} \bar{\lambda}_j z_j / n$, where n is the cardinality of E; then $||p_E||_\infty = 1$ and $\bigcap_{E \text{ finite} \subset \theta} \{\xi \in \bar{U}^n : |p_E(\xi)| = 1\} = \{\lambda\}$. □

We will use positive integers n for θ in the symbols Z_+^θ, P_θ, A_θ, $\mathbb{T}^\theta$, $\bar{U}^\theta$ to indicate the index set $\{1,2,\cdots,n\}$.

1.7 Definition: A Q-algebra is the quotient algebra of a function algebra with an identity modulo a closed ideal. Thus a Q-algebra is a commutative Banach algebra under the quotient norm, and has an identity.

1.8 Lemma: Any quotient algebra (modulo a closed ideal) of a Q-algebra is itself a Q-algebra.

Proof. Let A be a function algebra, $1 \in A$, and let I be a closed ideal. Let J be a closed ideal of A/I (a Q-algebra), then $(A/I)/J \cong A/J_1$, where $J_1 = \{f \in A : f+I \in J\}$. (We use the symbol "$\cong$" between Banach algebras or spaces to denote isometric isomorphism.) To prove this, one merely notes that quotient maps are onto, so A maps canonically onto $(A/I)/J$. Equality of the quotient norms (J and J_1) is also straightforwardly checked. □

1.9 <u>Lemma</u>: <u>Let</u> B <u>be</u> <u>a</u> <u>Q-algebra</u>, <u>let</u> $\{f_j : j \in \theta\}$ (θ <u>an index set</u>) <u>be a subset of the unit ball of</u> B (<u>that is</u>, $||f_j|| \leq 1$ <u>each</u> $j \in \theta$) <u>then there exists a bounded homomorphism</u> π <u>of</u> A_θ <u>into</u> B <u>such that</u>

$$\pi(\Sigma_{\alpha \in Z_+^\theta} a_\alpha z^\alpha) = \Sigma_\alpha a_\alpha f^\alpha \text{ for polynomials (elements of } P_\theta),$$

<u>and</u> $||\pi g|| \leq ||g||_\infty$, $g \in A_\theta$ (<u>recall</u> $f^\alpha = \Pi_{j\in\theta} f_j^{\alpha_j}, \alpha \in Z_+^\theta$). <u>Also</u> $\pi 1 = 1$.

Proof. Let $p \in P_\theta$ with $p = \Sigma_\alpha a_\alpha z^\alpha$, then $\pi p = \Sigma_\alpha a_\alpha f^\alpha$ is defined as an element of B, and π is clearly a homomorphism of P_θ into B. It remains to prove that $||\pi(p)|| \leq ||p||_\infty$ for each $p \in P_\theta$. Write $B = A/I$, where I is a closed ideal in the function algebra A, and $1 \in A$. Fix a positive number $r < 1$, then for each $j \in \theta$, there exists $g_{rj} \in A$ with $||g_{rj}||_A \leq 1$ and $rf_j = g_{rj} + I$.

Fix a polynomial $p = \Sigma_\alpha a_\alpha z^\alpha$ in P_θ, let $\psi \in M_A$, and let $h_r = \Sigma_\alpha a_\alpha g_r^\alpha$. Then $\psi(h_r) = \Sigma_\alpha a_\alpha \lambda^\alpha$, where $\lambda_j = \psi(g_{rj})$ and so $\lambda \in \overline{U}^\theta$. By 1.6 we have $|\psi(h_r)| \leq ||p||_\infty$, but $||h_r||_A = \sup\{|\psi(h_r)| : \psi \in M_A\}$ and so $||h_r||_A \leq ||p||_\infty$. Now map h_r into A/I and observe that $h_r + I \to \pi(p)$ in the norm as $r \to 1$ (since $h_r + I$ is a finite expression in $g_{rj} + I$). Hence $||\pi(p)||_B \leq \limsup_{r\to 1} ||h_r||_A \leq ||p||_\infty$. □

Lemma 1.9 is attributed to Craw in Davie [1]. In order to investigate the second dual of a Q-algebra we need some lemmas about weak topologies on Banach algebras. In the following two lemmas A denotes a commutative Banach algebra, and A* denotes its dual space. Thus A* is an A-module, with the module action defined by $(f\cdot\phi)(g) = \phi(fg)$, $(\phi \in A^*, f, g \in A)$, and $||f\cdot\phi|| \leq ||f||\ ||\phi||$. Let M be a closed subspace of A*, and let

τ be the weak topology $\sigma(A,M)$, that is, a net $\{f_\alpha\} \subset A$ τ-converges to $f \in A$ if and only if $\phi(f_\alpha) \overset{\alpha}{\to} \phi(f)$, $(\phi \in M)$.

1.10 Lemma: Multiplication in A is separately τ-continuous if and only if M is an A-submodule of A^*.

Proof. If multiplication is separately τ-continuous, $\phi \in M$, $f \in A$ then the functional $f\cdot\phi : g \mapsto \phi(fg)$ $(g \in A)$ is τ-continuous and thus $f\cdot\phi \in M$ (Rudin [2, p. 62]). Conversely, suppose M is an A-module, and let a net $\{f_\alpha\}$ τ-converge to $f \in A$. For $\phi \in M$, $g \in A$ we have $\phi(f_\alpha g) = (g\cdot\phi)(f_\alpha) \overset{\alpha}{\to} (g\cdot\phi)(f) = \phi(fg)$, (since $g\cdot\phi \in M$) and thus $f_\alpha g \overset{\alpha}{\to} fg$ in the τ-topology. □

1.11 Lemma: Let M be a closed submodule of A^*, and let I be a subspace of A. Then I is a τ-closed ideal in A if and only if $I = X^\perp$ and $X = I^\perp$ where X is a $\sigma(M,A)$-closed submodule of M (note $X^\perp = \{f \in A : \phi(f) = 0, (\phi \in X)\}$, and $I^\perp = \{\phi \in M : \phi(I) = 0\}$). Further in this case, i) A/I has an induced τ-topology in which multiplication is separately continuous; and $I^\perp$ is an A/I-module; ii) A/I is isomorphic to the space of $\sigma(M,A)$-continuous linear functionals on $I^\perp$; iii) $I^\perp$ is the space of τ-continuous linear functionals on A/I.

Proof. By the Hahn-Banach theorem (Rudin [2, p. 60]) $I = I^{\perp\perp}$ if and only if I is τ-closed, and $X = X^{\perp\perp}$ if and only if X is $\sigma(M,A)$-closed. Statements ii) and iii) are also standard arguments. Now suppose that I is an ideal. Let $\phi \in I^\perp$, $g \in A$, then $(g\cdot\phi)(f) = \phi(fg) = 0$ $(f \in I)$, thus $g\cdot\phi \in I^\perp$ and so $I^\perp$ is a submodule of M. Conversely suppose X is a submodule of M then $X^\perp$ is a τ-closed ideal in A. Indeed let $f \in X^{\perp}$, $g \in A$, then $\phi(fg) = (g\cdot\phi)(f) = 0$, $(\phi \in X)$, thus $fg \in X^{\perp}$. Statement i) follows from I being τ-closed and from Lemma 1.10. □

We now return to Q-algebras.

1.12 Lemma: *Let A be a function algebra, then the second dual, A**, is a function algebra and is realized as a weak-* closed subalgebra of a commutative W*-algebra (see 1.2.1). Further the dual, A*, is both an A and an A**-module. There exists a canonical homomorphism* $j: A \to A^{**}$.

Proof. Embed A isometrically into $C_o(X)$ for some locally compact Hausdorff space X, then $C_o(X)^* = M(X)$ and $M(X)^*$ is a commutative W*-algebra. Further M(X) is an $M(X)^*$-module (1.2.11). Then A* may be identified with $M(X)/A^{\perp}$ (where $A^{\perp} \subset M(X)$), and A** with $A^{\perp\perp} \subset M(X)^*$. Multiplication in $M(X)^*$ is separately weak-* continuous thus $A^{\perp\perp}$ is a norm and weak-* closed subalgebra of $M(X)^*$. By restriction M(X) is an $A^{\perp\perp}$-module, and this induces an action on $M(X)/A^{\perp}$ (since $A^{\perp\perp} \cdot A^{\perp} = \{0\}$). □

1.13 Corollary: *Let* τ *denote the weak-* topology on A** (that is,* $\sigma(A^{**}, A^*)$*), then multiplication in A** is separately* τ*-continuous.*

Proof. Use Lemma 1.10 on the algebra A** with M = A*. □

1.14 Theorem: *The second dual B** of a Q-algebra B is a Q-algebra, and multiplication is separately* $\sigma(B^{**},B^*)$ *continuous.*

Proof. Write $B = A/I$, where A is a function algebra with identity and I is a closed ideal in A. The dual of B may be identified with $I^{\perp}$ (where $I^{\perp} \subset A^*$) and $B^{**} = A^{**}/I^{\perp\perp}$, where $I^{\perp\perp} \subset A^{**}$ is the τ-closure of jI. By Lemma 1.12, A^{**} is a function algebra with identity, and multiplication is separately τ-continuous, thus $I^{\perp\perp}$ is a τ-closed ideal in A^{**}. Hence $B^{**} = A^{**}/I^{\perp\perp}$ is a Q-algebra. Lemma 1.11 shows that multiplication in B^{**} is separately $\sigma(B^{**},B^*)$-continuous. □

We will use τ also to denote the $\sigma(B^{**},B^*)$-topology (an extension of the previous definition).

1.15 <u>Corollary</u>: <u>Let</u> B <u>be</u> <u>a</u> Q-<u>algebra</u> <u>and</u> <u>let</u> Y <u>be</u> <u>a</u> <u>closed</u> B-<u>submodule</u> <u>of</u> B^*, <u>then</u>

i) $B^{**}/Y^{\perp}$ <u>is</u> <u>a</u> Q-<u>algebra</u>

ii) $B^{**}/Y^{\perp} \cong Y^*$

iii) <u>multiplication</u> <u>in</u> $B^{**}/Y^{\perp}$ <u>is</u> <u>separately</u> τ-<u>continuous, where</u> τ <u>is the induced</u> τ-<u>topology</u> <u>and</u> <u>is</u> <u>also</u> <u>equal</u> <u>to</u> $\sigma(B^{**}/Y^{\perp},Y)$.

Proof. To show that Y is a B^{**}-module, let $f \in B^{**}, \phi \in Y$. Then there exists a net $\{f_\alpha\} \subset B$ such that $jf_\alpha \overset{\alpha}{\to} f$ in τ. For each $g \in B^{**}$, $(f\cdot\phi)(g) = \phi(fg) = \lim_\alpha \phi((jf_\alpha)g) = \lim_\alpha (f_\alpha\cdot\phi)(g)$. But each $f_\alpha\cdot\phi \in Y$ and Y is norm hence $\sigma(B^*,B^{**})$-closed and so $f\cdot\phi \in Y$. Lemma 1.11 shows that $Y^{\perp}$ is a τ-closed ideal in B^{**}, and Lemma 1.8 shows that $B^{**}/Y^{\perp}$ is a Q-algebra. Statements ii) and iii) follow from Lemma 1.11. □

We have shown a method for constructing certain commutative Banach algebras which are dual spaces and have separately weak-*-continuous multiplication. By the Alaoglu theorem, the unit ball in such an algebra is a compact commutative semitopological semigroup (under multiplication). Further our examples need not be semi-simple, so the unit ball need not be separative (Chapter 4) as a discrete semigroup. Thus we have a class of semigroups, more general than the unit balls of L^∞-spaces, available for representations of semitopological semigroups. We now give an abstract definition of these special algebras.

1.16 Definition: A dual Q-algebra B is a Q-algebra which is the dual space of a Banach space B_* and in which multiplication is separately $\sigma(B,B_*)$-continuous.

Corollary 1.15 shows that the second dual B** of a Q-algebra is a dual Q-algebra, and so are the quotients of B** modulo τ-closed ideals. Every dual Q-algebra arises this way, since B may be identified with $B^{**}/B_*^{\perp}$, interpreting B_* as a closed B-submodule of B*.

§2. Q-representations and the algebra RQ(S)

Throughout this section S denotes a commutative semitopological semigroup with an identity 1. The semigroup S made discrete is denoted by S_d. The key idea is that if S is represented in the ball of a Q-algebra B then any homomorphism of Z_+^θ into S naturally induces a bounded homomorphism of A_θ into B (see Lemma 1.9). Indeed to study such representations of S it suffices to consider quotients of A_θ for sufficiently large θ. We will define the important algebra RQ, and prove a definitive theorem about the action on RQ of semigroup homomorphisms with dense range.

6.2.1

To draw a comparison with Chapter 2, $R(S)$ is a Banach algebra of continuous functions on S whose dual $A(S)$ is a dual function algebra in whose ball S is weak-* continuously represented. The algebra $RQ(S)$ has similar properties except that $RQ(S)^*$ is a dual Q-algebra (Theorem 2.10). We expect that $RQ(S)$ separates points on a much larger class of semigroups than $R(S)$ does. This question will be investigated in Chapter 7.

2.1 Definition: A Q-representation of S is a homomorphism ψ of S into the unit ball of a dual Q-algebra B such that $\psi 1 = 1$ and which is weak-* continuous, that is $x_\alpha \overset{\alpha}{\to} x$ in S implies $\psi x_\alpha \overset{\alpha}{\to} \psi x$ in $\sigma(B,B_*)$. Let ψ^* denote the bounded linear map of B_* into $C^B(S)$ given by $\psi^*\phi(x) = \phi(\psi x)$ $(\phi \in B_*, x \in S)$.

2.2 Theorem: Let ψ be a Q-representation of S in a dual Q-algebra B, then $\psi^*\phi \in WAP(S)$ for each $\phi \in B_*$.

Proof. Note that B_* is a B-module and for $\psi \in B_*$ the set $O(\phi) = \{f\cdot\phi : f \in \text{unit ball } B\}$ is $\sigma(B_*,B)$-compact. Indeed, let $\{f_\alpha\}$ be a net in the unit ball of B converging weak-* to $f \in B$, then $(f_\alpha\cdot\phi)(g) = (g\cdot\phi)(f_\alpha) \overset{\alpha}{\to} (g\cdot\phi)(f) = (f\cdot\phi)(g)$ $(g \in B)$. Thus the map $f \mapsto f\cdot\phi$ is $\sigma(B,B_*)$ to $\sigma(B_*,B)$ continuous. The unit ball in B is $\sigma(B,B_*)$ compact and thus $O(\phi)$ is $\sigma(B_*,B)$-compact. For $x \in S$ we have $R(x)\psi^*\phi = \psi^*(\psi(x)\cdot\phi)$. Thus the orbit $O(\psi^*\phi) = \{R(x)\psi^*\phi : x \in S\} \subset \psi^*(O(\phi))$, but ψ^* is weakly continuous $B_* \to C^B(S)$ and so $\psi^*\phi$ is weakly almost periodic. □

2.3 Definition: Let $RQ(S)$ be the set of $f \in C^B(S)$ for which

(*) $$|\Sigma_{\alpha\in Z^n_+} a_\alpha f(x^\alpha)| \leq K_f||\Sigma_{\alpha\in Z^n_+} a_\alpha z^\alpha||_\infty$$

for each $n = 1,2,\cdots$, $x_1,\cdots,x_n \in S$, and $\Sigma_{\alpha\in Z^n_+} a_\alpha z^\alpha \in P_n$ (see Definition 1.5, thus only finitely many $a_\alpha \neq 0$) some $K_f < \infty$. For $f \in RQ(S)$ let $||f||_Q$ denote the least value of K_f for which (*) holds.

2.4 _Proposition_: _RQ(S) with the_ $||\cdot||_Q$_-norm is a normed linear space of continuous bounded functions on S, and_ $\sup_S|f| \leq ||f||_Q$.

Proof. Let $x \in S$, then $|f(x)| \leq ||f||_Q||z_1||_\infty = ||f||_Q$. Condition (*) shows directly that $||f+g||_Q \leq ||f||_Q + ||g||_Q$ and that $||cf||_Q = |c|\ ||f||_Q$, for $f,g \in RQ(S)$, $c \in \mathbb{C}$. Thus $RQ(S)$ is a normed linear space. □

2.5 _Theorem_: _Let_ ψ _be a Q-representation of S in the dual Q-algebra B. Then_ ψ^* _maps_ B_* _into RQ(S) and_ $||\psi^*\phi||_Q \leq ||\phi||$ $(\phi \in B_*)$.

Proof. Let $\phi \in B_*$ then $\psi^*\phi \in C^B(S)$. For $n = 1,2,\cdots$, points $x_1,x_2,\cdots,x_n \in S$ define the homomorphism $\pi: Z^n_+ \to S$ by $\pi\alpha = x^\alpha$ $(\alpha \in Z^n_+)$. Then $\psi\circ\pi$ maps Z^n_+ into the unit ball of B, a Q-algebra. By Lemma 1.9, $||\Sigma_{\alpha\in Z^n_+} a_\alpha\psi\circ\pi(\alpha)||_B \leq ||\Sigma_\alpha a_\alpha z^\alpha||_\infty$ for each polynomial $\Sigma_\alpha a_\alpha z^\alpha \in P_n$. Thus $|\Sigma_\alpha a_\alpha\psi^*\phi(x^\alpha)| = |\phi(\Sigma_\alpha a_\alpha\psi(\pi\alpha))|$ $\leq ||\phi||\ ||\Sigma_\alpha a_\alpha\psi(\pi\alpha)||_B \leq ||\phi||\ ||\Sigma_\alpha a_\alpha z^\alpha||_\infty$, and so $\psi^*\phi \in RQ(S)$ with $||\psi^*\phi||_Q \leq ||\phi||$. □

2.6 _Corollary_: $R(S) \subset RQ(S)$, _and_ $||f||_Q \leq ||f||_R$ _for_ $f \in R(S)$.

Proof. Recall that the dual of $R(S)$, $A(S)$, is a function algebra and ρ is a weak-* continuous representation of S in the unit ball of $A(S)$. However $A(S)$ is a dual Q-algebra and $\rho^* f(x) = f(x)$ $(f \in R(S),\ x \in S)$ so ρ^* maps $A(S)_* = R(S)$ into $RQ(S)$ (by Theorem 2.5) and $||f||_Q \leq ||f||_R$. □

2.7 <u>Theorem: RQ(S) is a normed algebra and is conjugate closed and translation-invariant.</u>

Proof. Let $f_1, f_2 \in RQ(S)$ and fix points $x_1, \cdots, x_n \in S$. Condition 2.3(*) shows that f_1, f_2 define bounded linear functionals ϕ_1, ϕ_2 respectively on A_n, given by $\phi_j(\Sigma_{\alpha \in Z^n_+} a_\alpha z^\alpha) = \Sigma_\alpha a_\alpha f_j(x^\alpha)$ for polynomials $\Sigma_\alpha a_\alpha z^\alpha$, and $||\phi_j|| \leq ||f_j||_Q$, $(j = 1,2)$. By the Hahn-Banach and Riesz theorems there exist measures $\mu_1, \mu_2 \in M(\mathbb{T}^n)$ so that $\phi_j(g) = \int_{\mathbb{T}^n} g\, d\mu_j$ $(g \in A_n)$, and $||\mu_j|| = ||\phi_j||$ $(j = 1,2)$. In particular $f_j(x^\alpha) = \phi_j(z^\alpha) = \int_{\mathbb{T}^n} z^\alpha d\mu_j = (\mu_j)^\wedge(\alpha)$. Let $\mu = \mu_1 * \mu_2$ (convolution in $M(\mathbb{T}^n)$) then $||\mu|| \leq ||\mu_1||\ ||\mu_2||$ and $\hat{\mu}(\alpha) = \hat{\mu}_1(\alpha)\hat{\mu}_2(\alpha)$, each $\alpha \in Z^n_+$. For any polynomial we have $|\Sigma_\alpha a_\alpha f_1(x^\alpha) f_2(x^\alpha)| = |\Sigma_\alpha a_\alpha \hat{\mu}_1(\alpha)\hat{\mu}_2(\alpha)| = |\Sigma_\alpha a_\alpha \hat{\mu}(\alpha)| =$
$= |\int_{\mathbb{T}^n} \Sigma_\alpha a_\alpha z^\alpha d\mu| \leq ||\Sigma_\alpha a_\alpha z^\alpha||_\infty ||\mu|| \leq ||\Sigma_\alpha a_\alpha z^\alpha||_\infty ||\mu_1||\ ||\mu_2||$
$\leq ||\Sigma_\alpha a_\alpha z^\alpha||_\infty ||f_1||_Q ||f_2||_Q$. But $x_1, \cdots, x_n$ were arbitrary, thus $f_1 f_2 \in RQ(S)$ and $||f_1 f_2||_Q \leq ||f_1||_Q ||f_2||_Q$. (It is clear that $f_1 f_2 \in C^B(S)$).

To show that $RQ(S)$ is conjugate-closed let $f \in RQ(S)$ and $x_1, \cdots, x_n \in S$. Mimic the above construction to obtain $\mu \in M(\mathbb{T}^n)$ such that $||\mu|| \leq ||f||_Q$ and $\hat{\mu}(\alpha) = f(x^\alpha)$ $(\alpha \in Z^n_+)$. Let $\nu \in M(\mathbb{T}^n)$ be defined by $\int_{\mathbb{T}^n} g\, d\nu = (\int_{\mathbb{T}^n} \overline{g(x^{-1})}\, d\mu(x))^-$ then $||\nu|| = ||\mu||$

and $\hat{\nu}(\alpha) = \overline{\hat{\mu}(\alpha)}$. Thus for any polynomial in P_n
$|\Sigma_\alpha a_\alpha \overline{f}(x^\alpha)| = |\Sigma_\alpha a_\alpha \overline{\hat{\mu}(\alpha)}| = |\Sigma_\alpha a_\alpha \hat{\nu}(\alpha)| \leq ||\Sigma_\alpha a_\alpha z^\alpha||_\infty ||\nu||$
$= ||\Sigma_\alpha a_\alpha z^\alpha||_\infty ||\mu|| \leq ||\Sigma_\alpha a_\alpha z^\alpha||_\infty ||f||_Q$. Thus $\overline{f} \in RQ(S)$ and $||\overline{f}||_Q = ||f||_Q$ (by symmetry). For translation invariance, let $f \in RQ(S)$, $y, x_1, \cdots, x_n \in S$. For a polynomial $\Sigma_\alpha a_\alpha z^\alpha \in P_n$ we have $|\Sigma_\alpha a_\alpha (R(y)f)(x^\alpha)| = |\Sigma_\alpha a_\alpha f(yx^\alpha)| \leq ||f||_Q ||z_{n+1} \Sigma_{\alpha \in Z_+^n} a_\alpha z^\alpha||_\infty$
$= ||f||_Q ||\Sigma_\alpha a_\alpha z^\alpha||_\infty$. The idea is to set $x_{n+1} = y$ and apply condition (*) to the polynomials $z_{n+1} \Sigma_{\alpha \in Z_+^n} a_\alpha z^\alpha$. Since S is semitopological $R(y)f \in C^B(S)$ and so $R(y)f \in RQ(S)$ with $||R(y)f||_Q \leq ||f||_Q$. □

We now show the basic result that if S is discrete then RQ(S) is isomorphic to the dual of a quotient of A_θ, some θ. Thus RQ(S)* is a dual Q-algebra (Theorem 1.14). This also shows that RQ(S) is a Banach space. If S is semitopological, then RQ(S) is a translation-invariant closed subspace of $RQ(S_d)$ and we will use Corollary 1.15 to show that RQ(S)* is a dual Q-algebra.

2.8 <u>Theorem</u>: <u>Let</u> S <u>be discrete and let</u> $\{x_j : j \in \theta\}$ (θ <u>some index set</u>) <u>be a set of generators for</u> S. <u>Then</u> $RQ(S) \cong (A_\theta/I)^*$ <u>for a certain closed ideal</u> I <u>in</u> A_θ. <u>Closed</u> A_θ/I<u>-submodules of</u> $(A_\theta/I)^*$ <u>correspond to translation-invariant closed subspaces of</u> RQ(S).

Proof. We first construct a representation of S in a quotient of A_θ and then show that this representation is in a sense maximal. Define a homomorphism π of P_θ onto $c_c(S)$, $\pi(\Sigma_\alpha a_\alpha z^\alpha) = \Sigma_\alpha a_\alpha \delta(x^\alpha)$ (where $\delta(y)$ is the function which is 1 at y, 0 elsewhere). The map is onto since $\{x_j : j \in \theta\}$ generates S. Recall $c_c(S)$ is the semigroup algebra with the multiplication $\delta(u)\delta(v) = \delta(uv)$ $(u, v \in S)$.

The monomials z^α map onto the points of S which we identify with the functions $\delta(u) \in c_c(S)$.

Let I be the kernel of π, then we have a faithful representation of S by mapping $x^\alpha \mapsto z^\alpha + I_P \in P_\theta/I_P$ $(\alpha \in Z_+^\theta)$, (indeed $c_c(S) \cong P_\theta/I_P)$. Let I be the closure of I_P in A_θ thus I is a closed ideal and there exists a canonical homomorphism $P_\theta/I_P \to A_\theta/I$ (with dense range). Define the representation ψ of S by $\psi x^\alpha = z^\alpha + I \in A_\theta/I$. Note that $||\psi x^\alpha|| = ||z^\alpha+I||$ $\leq ||z^\alpha||_\infty = 1$, thus ψ maps S into the unit ball of A_θ/I, a Q-algebra.

We are ready to show that $RQ(S) \cong (A_\theta/I)^*$. The argument of Theorem 2.5 shows that for $\phi \in (A_\theta/I)^*$ the function f on S defined by $f(x^\alpha) = \phi(z^\alpha+I)$ $(\alpha \in Z_+^\theta)$ is in RQ(S) (S is discrete!) and $||f||_Q \leq ||\phi||$. Given $\varepsilon > 0$, by the definition of $||\phi||$ there exists a polynomial $p = \Sigma_\alpha a_\alpha z^\alpha$ with $||p||_\infty = 1$ and $|\phi(p+I)| > ||\phi|| - \varepsilon$. Then $|\Sigma_\alpha a_\alpha f(x^\alpha)| = |\Sigma_\alpha a_\alpha \phi(z^\alpha+I)|$ $= |\phi(\Sigma_\alpha a_\alpha z^\alpha+I)| > ||\phi|| - \varepsilon$, hence $||f||_Q \geq ||\phi|| - \varepsilon$. It remains to show that if $f \in RQ(S)$ then f corresponds to some element of $(A_\theta/I)^*$. Let $f \in RQ(S)$, $\Sigma_\alpha a_\alpha z^\alpha \in P_\theta$, and define the linear functional ϕ by $\phi(\Sigma_\alpha a_\alpha z^\alpha) = \Sigma_\alpha a_\alpha f(x^\alpha)$. Condition (*) shows that ϕ extends to a bounded linear functional on A_θ with $||\phi|| \leq ||f||_Q$. We observe that ϕ annihilates the kernel of π, namely I_P, and thus $\phi(I) = 0$ (ϕ is continuous). This shows that $\phi \in I^\perp \cong (A_\theta/I)^*$. Let $f \in RQ(S)$ correspond to $\phi \in I^\perp \subset A_\theta^*$ as above. Let $y \in S$ and write $y = x^\beta$ for some $\beta \in Z_+^\theta$. Let ϕ_y be the element of I corresponding to $R(y)f$, thus $\phi_y(\Sigma_\alpha a_\alpha z^\alpha) = \Sigma_\alpha a_\alpha R(y)f(x^\alpha) = \Sigma_\alpha a_\alpha f(x^{\alpha+\beta}) = \phi(z^\beta \Sigma_\alpha a_\alpha z^\alpha)$ $= (z^\beta\cdot\phi)(\Sigma_\alpha a_\alpha z^\alpha)$ for each $\Sigma_\alpha a_\alpha z^\alpha \in P_\theta$. Let X be a closed A_θ-

submodule of $I^{\perp}$, then $\phi \in X$ implies $\phi_y \in X$ for each $y \in S$ (since $\phi_y = z^{\beta}\cdot\phi$, $y = x^{\beta}$, $\beta \in Z_+^{\theta}$), thus X determines a closed translation-invariant subspace of RQ(S). Conversely, let $X \subset I^{\perp}$ correspond to a translation-invariant closed subspace of RQ(S), then for each $\phi \in X$, $\beta \in Z_+^{\theta}$ we have $z^{\beta}\cdot\phi \in X$. Thus $\phi \in X$, $p \in P_{\theta}$ implies $p\cdot\phi \in X$, and the density of P_{θ} in A_{θ} and X being closed shows that X is an A_{θ}- , hence an A_{θ}/I-submodule of $I^{\perp}$ (see 1.11). □

The idea of the above proof will be referred to again in Chapter 7.

2.9 Corollary: RQ(S) _is a commutative Banach algebra_ (S _semi-topological_).

Proof. It remains to show that RQ(S) is complete. The theorem shows that $RQ(S_d)$ is a Banach space. We claim that RQ(S) is a closed subspace of $RQ(S_d)$ since convergence in the Q-norm implies uniform convergence (sup-norm, see Proposition 2.4). □

2.10 Theorem: RQ(S)* _is a dual_ Q-_algebra, and there exists a canonical_ Q-_representation_ σ _of_ S _in the unit ball of_ RQ(S)*, _given by_ $(\sigma x)(f) = f(x)$ $(x \in S,\ f \in RQ(S))$.

Proof. Theorems 1.14 and 2.8 show that $RQ(S_d)^*$ is a dual Q-algebra. We see that RQ(S) is a translation-invariant closed subspace (see Corollary 2.9) of $RQ(S_d)$, and thus a closed submodule.

By Corollary 1.15, RQ(S)* is a dual Q-algebra (a quotient of $RQ(S_d)^*$). The argument of Theorem 2.8 shows that multiplication in $c_c(S_d)$ corresponds to multiplication in A_{θ}/I, hence in $(A_{\theta}/I)^{**}$, thus $\sigma: S \to RQ(S)^*$ is a homomorphism. It is τ-continuous, for if a net $\{x_\lambda\} \subset S$ converges to $x \in S$ and $f \in RQ(S)$ then $(\sigma x_\lambda)(f) = f(x_\lambda) \overset{\lambda}{\to} f(x) = (\sigma x)(f)$, since $RQ(S) \subset C^B(S)$. □

2.11 Corollary: $RQ(S) \subset WAP(S)$.

Proof. Theorems 2.2 and 2.10 imply the result. □

2.12 Theorem: Let X be a closed translation-invariant subspace of RQ(S), then there exists a Q-representation of S in a dual Q-algebra B such that ψ^* maps B_* isometrically onto X.

Proof. As in Theorem 2.8, X is a closed submodule of RQ(S), and Corollary 1.15 shows that $RQ(S)^*/X^\perp$ is a dual Q-algebra. Let $B = RQ(S)^*/X^\perp$, then $B_* \cong X$, and ψ^* realizes the isometric isomorphism. □

The definition of RQ(S) makes it easy to prove a theorem about pointwise limits of functions in RQ(S).

2.13 Theorem: Let $\{f_\lambda\}$ be a net in RQ(S) such that $f_\lambda \overset{\lambda}{\to} f \in C^B(S)$ pointwise and $||f_\lambda||_Q \leq K < \infty$, for all λ, then $f \in RQ(S)$ and $||f||_Q \leq K$.

Proof. For each set $x_1, \cdots, x_n \in S$ we have $|\Sigma_{\alpha \in Z_+^n} a_\alpha f_\lambda(x^\alpha)| \leq K||\Sigma_\alpha a_\alpha z^\alpha||_\infty$ for each $\Sigma_\alpha a_\alpha z^\alpha \in P_n$. The left side involves only finitely many points in S, so it converges to $|\Sigma_{\alpha \in Z_+^n} a_\alpha f(x^\alpha)|$. Thus f satisfies condition (2.3*) with $||f||_Q \leq K$, and $f \in C^B(S)$, hence $f \in RQ(S)$. □

In Chapter 3 we gave a proof that $R(S)$ is closed under such limits provided that S has a dense inverse semigroup. Corollary

2.6 and Theorem 2.13 show that in any case a norm-bounded pointwise convergent net in $R(S)$ converges to an element of $RQ(S_d)$ (note $RQ(S) = RQ(S_d) \cap C^B(S)$).

We now investigate the effect of semigroup homomorphisms on RQ. The situation is quite satisfactory. Indeed it is possible to determine $RQ(S_0)$ if S_0 is a quotient of S, and RQ(S) is known. Further to define a function in RQ(S) it suffices for condition 2.3* to hold on a dense subsemigroup, which is surprising since the map $(x_1,\cdots,x_n) \mapsto x^\alpha$ some $\alpha \in Z_+^n$ need not even be separately continuous (Remark 3.1.10).

2.14 <u>Proposition</u>: <u>Let</u> π <u>be a continuous homomorphism of</u> S_1 <u>into</u> S_2, <u>such that</u> $\pi 1 = 1$, <u>then</u> π <u>induces a homomorphism</u> $\pi^\#: RQ(S_2) \to RQ(S_1)$ <u>such that</u> $||\pi^\# f||_Q \le ||f||_Q$ and $\pi^\# f(x) = f(\pi x)$ $(f \in RQ(S_2),\ x \in S_1)$.

Proof. It is clear that $\pi^\# f \in C^B(S_1)$ for $f \in RQ(S_2)$. For each $x_1,\cdots,x_n \in S$, $\pi(x^\alpha) = (\pi x)^\alpha$ $(\alpha \in Z_+^n)$ and it is clear that $\pi^\# f$ satisfies condition 2.3* with some constant $\le ||f||_Q$. □

2.15 <u>Theorem</u>: <u>Let</u> π <u>be a continuous homomorphism of</u> S_1 <u>onto a dense subsemigroup of</u> S_2, <u>such that</u> $\pi 1 = 1$, <u>then</u> $RQ(S_2) \cong \pi^\# RQ(S_2) = RQ(S_1) \cap \pi^\# C^B(S_2)$.

Proof. Since πS_1 is dense in S_2 we see that $\pi^\#$ is one-to-one on $RQ(S_2)$. Further $\pi^\# RQ(S_2) \subset RQ(S_1) \cap \pi^\# C^B(S_2)$, by 2.14, so we must show that if $f \in C^B(S_2)$ and $\pi^\# f \in RQ(S_1)$ then $f \in RQ(S_2)$ and $||f||_Q \le ||\pi^\# f||_Q$ (hence "=").

We claim that $X = RQ(S_1) \cap \pi^\# C^B(S_2)$ is a closed translation-invariant subspace of $RQ(S_1)$. For $f \in C^B(S_2)$, $x \in S_1$ we have $R(x)\pi^{\#}f = \pi^{\#}(R(\pi x)f)$, and we get the translation-invariance. Further $\pi^{\#}$ is an isometry of $C^B(S_2)$ into $C^B(S_1)$ because $\sup_{S_1}|\pi^{\#}f| = \sup_{\pi S_1}|f| = \sup_{S_2}|f|$ since πS_1 is dense in S_2. Thus X is closed (recall $\sup_{S_1}|\pi^{\#}f| \leq ||\pi^{\#}f||_Q$). By Theorem 2.12 there exists a Q-representation ψ of S_1 in the unit ball of a dual Q-algebra B, and $\psi^*B_* \cong X$. We will show that ψ induces a Q-representation of S_2.

For each $y \in S_2$ define $\psi'y$ to be the bounded linear functional on X given by $\psi'y(\pi^{\#}f) = f(y)$, ($f \in C^B(S_2)$ and $\pi^{\#}f \in RQ(S_1)$). Note that $|f(y)| \leq \sup_{S_2}|f| = \sup_{S_1}|\pi^{\#}f| \leq ||\pi^{\#}f||_Q$, thus $\psi'y$ is in the unit ball of B ($\cong X^*$). Further ψ' is τ-continuous, and for $x \in S_1$ it is clear that $\psi'(\pi x) = \psi x$. Thus ψ' is multiplicative on a dense subsemigroup of S_2 and is hence a homomorphism of S_2.

Let $f \in C^B(S_2)$ and $\pi^{\#}f \in RQ(S_1)$, then by Theorem 2.5 applied to ψ' the function $f \in RQ(S_2)$ and $||f||_Q \leq ||\pi^{\#}f||_X$, where $||\cdot||_X$ is the norm on X, that is, the $RQ(S_2)$ norm. Finally $||f||_Q \leq ||\pi^{\#}f||_Q \leq ||f||_Q$. □

2.16 Corollary: Let S_0 be a dense subsemigroup of S such that $1 \in S_0$ and let $f \in C^B(S)$. Then $f \in RQ(S)$ if and only if f satisfies 2.3(*) for all subsets $\{x_1,\cdots,x_n\} \subset S_0$, and $||f||_Q = ||f|S_0||_Q$.

We point out that theorems like the above for $R(S)$ were proved in Chapter 2 under the hypothesis of a dense inverse semigroup.

A quotient semigroup S_0 of S is a semigroup of subsets of S, so that S_0 is semitopological under the induced topology. Let

π be the canonical homomorphism: $S \to S_0$ then $f \in \pi^{\#}C^B(S_0)$ if and only if $f \in C^B(S)$ and $\pi x = \pi y$ implies $f(x) = f(y)$ $(x,y \in S)$.

2.17 Corollary: Let S_0 be a quotient semigroup of S, then $RQ(S_0) \cong \{f \in RQ(S) : \pi x = \pi y$ implies $f(x) = f(y),\ (x,y \in S)\}$.

This Corollary will be the key tool in finding $RQ(\mathbb{R}_1)$, where $\mathbb{R}_1$ is the nil-thread, the interval $[0,1]$ with $xy = \min(1,x+y)$.

§3. Sources and related work

The definition of RQ(S) and Theorems 2.7, 2.8, 2.10 are due to Dunkl and were announced by him in [1].

Basic references for the polydisc algebra are Rudin's book [3] and Stout's book [1].

Arens [1] and Civin and Yood [1] have investigated the second duals of Banach algebras.

Chapter 7. Special cases of Q-representations

We continue our study of Q-representations and RQ(S). We will pay close attention to discrete semigroups, semitopological semigroups with dense inverse semigroups (this includes idempotent semigroups and locally compact abelian groups), and the important special cases, Z_+^n, $\mathbb{R}_+^n$, and the nil-thread $\mathbb{R}_+/[1,\infty)$. Cole's theorem on the isomorphic isometric representation of a Q-algebra in the algebra of bounded operators on some Hilbert space is an important tool for investigating RQ(S) for a semigroup S of type $\mathcal{U}$. Some questions for further research appear at the end of the chapter.

§1. Discrete semigroups

We first recall some of the ideas from Theorem 6.2.8. Let S be a discrete commutative semigroup with unit 1, and let $\{x_j : j \in \theta\}$ (some index set θ) be a set of generators for S. For $x \in S$ let $\delta(x)$ denote the function which is 1 at x, 0 elsewhere, thus the set $\{\delta(x) : x \in S\}$ is a basis for $c_c(S)$. Also $c_c(S)$ is the semigroup algebra, with multiplication $\delta(xy) = \delta(x)\delta(y)$ $(x,y \in S)$. Define the homomorphism π of P_θ onto $c_c(S)$ by $\pi(\Sigma_\alpha a_\alpha z^\alpha) = \Sigma_\alpha a_\alpha \delta(x^\alpha)$ (finite sum). Let I_P be the kernel of π, then $c_c(S) = P_\theta / I_P$.

1.1 <u>Lemma</u>: I_P <u>is the linear span of</u> $\{z^\alpha - z^\beta : \alpha, \beta \in Z_+^\theta,\ \pi(z^\alpha - z^\beta) = 0\}$.

Proof. Let $p \in I_P$ with $p = \Sigma_\alpha a_\alpha z^\alpha$, thus $\Sigma_\alpha a_\alpha \delta(x^\alpha) = 0$. So for each $y \in S$, $\Sigma\{a_\alpha : \pi(z^\alpha) = \delta(x^\alpha) = \delta(y)\} = 0$. Fix $y \in S$ and let $\alpha_1,\cdots,\alpha_k \in Z_+^\theta$ such that $a_{\alpha_j} \neq 0$ and $\pi z^{\alpha_j} = \delta(y)$ $(j = 1,\cdots,k)$. Thus $\Sigma_{j=1}^k a_{\alpha_j} = 0$ and $\Sigma_{j=1}^k a_{\alpha_j} z^{\alpha_j}$ $= \Sigma_{j=2}^k a_{\alpha_j}(z^{\alpha_j}-z^{\alpha_1})$. But p is a finite linear combination of such terms, thus $I_p \subset \text{span}\ \{z^\alpha-z^\beta : x^\alpha = x^\beta\}$. The opposite inclusion is obvious. □

A relation on S is an equation of the sort $x^\alpha = x^\beta$ $(\alpha,\beta \in Z_+^\theta)$. Say S is _finitely presented_ if finitely many generators and relations determine S.

1.2 _Theorem_: _A finitely generated commutative semigroup is finitely presented_.

Proof. Let $\{x_1,\cdots,x_n\}$ generate S. Now I_p is an ideal in P_n so by the Hilbert basis theorem I_p is finitely generated. Let $g_1,\cdots,g_k$ generate I_p. By Lemma 1.1 each g_j is a finite linear combination of $\{z^\alpha-z^\beta : x^\alpha = x^\beta\}$. Thus finitely many polynomials of the form $z^\alpha-z^\beta$ where $x^\alpha = x^\beta$ generate I_p. But $c_c(S) = P_n/I_p$ so S is determined by finitely many relations $x^\alpha = x^\beta$. □

This theorem was stated and first proved by Rédei [1, p. 179], (see also Freyd [1]).
We continue as in Theorem 6.2.8. Let I be the closure of I_p in A_θ and set up the homomorphisms $c_c(S) \to P_\theta/I_p \to A_\theta/I$. Thus we obtain a representation ψ of S in the unit ball of A_θ/I given by $\psi x^\alpha = z^\alpha + I \in A_\theta/I$ $(\alpha \in Z_+^\theta)$.

1.3 _Definition_: The seminorm $||\cdot||_{QA}$ on $c_c(S)$ is given by $||\Sigma_{x\in S} a_x \delta(x)||_{QA} = ||\Sigma_{x\in S} a_x(\psi x)||_{A_\theta/I}$. Denote the completion of $c_c(S)$ in $||\cdot||_{QA}$ by $QA(S)$.

The following is implicit in Theorem 6.2.8.

1.4 Theorem: The space QA(S) is a Q-algebra isometrically isomorphic to A_θ/I and its dual space is RQ(S).

1.5 Corollary: Let S be semitopological and let $f \in C^B(S)$, then $f \in RQ(S)$ if and only if $|\Sigma_{x\in S} a_x f(x)| \leq K_f ||\Sigma_{x\in S} a_x \delta(x)||_{QA}$ $(\Sigma_{x\in S} a_x \delta(x) \in c_c(S))$, some constant $K_f < \infty$, and $||f||_Q$ is the least value of K_f for which the bound holds.

Whether RQ(S) separates points on S is an important question. The following three propositions give more insight into what is involved. An equivalent condition is that $\psi: S \to A_\theta/I$ be one-to-one, that is, $P_\theta \cap I = I_P$.

1.6 Proposition: Let $E \subset \overline{U}^\theta$ (the θ-fold closed unit disc) be the zero-set of I, then $E \cong \hat{S}$ under the identification $\psi^*\lambda(x^\alpha) = \lambda^\alpha$ where $\lambda \in E$, $\alpha \in Z_+^\theta$. Further E is the maximal ideal space of $A_\theta/I \cong QA(S)$.

Proof. Let $\chi \in \hat{S}$ and let $\lambda_j = \chi(x_j)$ $(j \in \theta)$, then $\lambda \in \overline{U}^\theta$ (i.e. $|\lambda_j| \leq 1$ all j). For any pair $\alpha,\beta \in Z_+^\theta$ such that $x^\alpha = x^\beta$ we have $\lambda^\alpha = \chi(x^\alpha) = \chi(x^\beta) = \lambda^\beta$ so the polynomial $z^\alpha - z^\beta$ vanishes at λ, but the span of these polynomials is dense in I by Lemma 1.1, hence each function in I vanishes at λ, that is, $\lambda \in E$, and $\psi^*\lambda = \chi$. Conversely, given $\lambda \in E$, the function $\psi^*\lambda(x^\alpha) = \lambda^\alpha$ $(\alpha \in Z_+^\theta)$ is well-defined on S (since $x^\alpha = x^\beta$ implies $\lambda^\alpha - \lambda^\beta = 0$) and is a semicharacter. Similar methods show that E is the maximal ideal space of A_θ/I. It is trivial to check that the isomorphism $\hat{S} \cong E$ is a homeomorphism. □

It will be convenient to identify $\hat{S}$ with E and thus consider $\hat{S}$ as a (compact) subset of $\overline{U}^{\theta}$.

The above proposition illustrates in another way that if S is separative (thus $\hat{S}$ separates the points of S) then RQ(S) separates the points.

1.7 Proposition: Let $\mu \in M(\mathbb{T}^{\theta})$, then $\mu \perp I$ if and only if $\hat{\mu}(\alpha) = \hat{\mu}(\beta)$ whenever $x^{\alpha} = x^{\beta}$ $(\alpha,\beta \in Z_{+}^{\theta})$.

Proof. Since I is the closure of I_P in $C(\mathbb{T}^{\theta})$ it is clear that $\mu \perp I$ if and only if $\mu \perp I_P$, that is, $\int_{\mathbb{T}^{\theta}}(z^{\alpha}-z^{\beta})\,d\mu = 0$ whenever $z^{\alpha}-z^{\beta} \in I_P$ (Lemma 1.1). □

Let $I^{\perp}$ denote the set of $\mu \in M(\mathbb{T}^{\theta})$ such that $\hat{\mu}(\alpha) = \hat{\mu}(\beta)$ whenever $x^{\alpha} = x^{\beta}$ $(\alpha,\beta \in Z_{+}^{\theta})$. From Theorem 1.4 we see that RQ(S) is isomorphic to the restriction of $I^{\perp}$ to A_{θ}. We point out that $I^{\perp}$ is closed under convolution (that is, RQ(S) is closed under pointwise multiplication). The following condition for RQ(S) to separate points is now clear:

1.8 Proposition: Let $\alpha,\beta \in Z_{+}^{\theta}$ with $x^{\alpha} \neq x^{\beta}$, then there exists $f \in RQ(S)$ such that $f(x^{\alpha}) \neq f(x^{\beta})$ if and only if there exists $\mu \in I^{\perp} \subset M(\mathbb{T}^{\theta})$ such that $\hat{\mu}(\alpha) \neq \hat{\mu}(\beta)$.

For example, if $\{\gamma \in Z_{+}^{\theta} : x^{\gamma} = x^{\alpha}\}$ is finite then there exists $f \in RQ(S)$ with $f(x^{\alpha}) = 1$ and $f(y) = 0$ for $y \neq x^{\alpha}$.

The general problem consists in finding measures μ on $\mathbb{T}^{\theta}$ such that $\hat{\mu}$ is constant on each equivalence class $\{\alpha \in Z_{+}^{\theta}: x^{\alpha} = x^{\alpha_0}$, some $\alpha_0 \in Z_{+}^{\theta}\}$ and separates two desired classes. Here are some infinite nonseparative examples (in Z_{+}^{θ} let e_j denote the element of Z_{+}^{θ} which is 1 in the $j^{\underline{th}}$ coordinate, otherwise zero):

1.9 Example: Let $S = \{1,0\} \cup \{x_j\}_{j\in\theta}$ (θ arbitrary) with $x_ix_j = x_i0 = 0$ $(i,j \in \theta)$, then $c_c(S) \subset RQ(S)$, (indeed $RQ(S) = \ell^\infty(S)$, see Dunkl [2]).

Proof. First $\delta(1)$ is a semicharacter, so $\delta(1) \in RQ(S)$. Represent S in A_θ/I, where I_p is generated by $z_iz_jz_k - z_iz_j$ (for all $i,j,k \in \theta$), thus I_p consists of all polynomials of degree ≥ 2. For any $j \in \theta$ there exists a measure $\mu_j \in M(\mathbb{T}^\theta)$ such that $\hat{\mu}(e_j) = 1$ and $\hat{\mu}(\alpha) = 0$ for all $\alpha \neq e_j$. Then $\mu \in I^\perp$ and so $\delta(x_j) \in RQ(S)$.

Define $\mu \in M(\mathbb{T}^\theta)$ by $\int_{\mathbb{T}^\theta} f d\mu = \frac{1}{2\pi}\int_{-\pi}^{\pi} f((e^{i\phi}))e^{-i\phi}d\phi$, (where $(e^{i\phi})$ denotes the element $\lambda \in \mathbb{T}^\theta$ with $\lambda_j = e^{i\phi}$ for all j and $f \in C(\mathbb{T}^\theta)$). Then $\hat{\mu}(e_j) = 1$ and $\hat{\mu}(\alpha) = 0$ for each $\alpha \in Z_+^\theta$ with $\alpha \neq e_j$, all j. Thus $\mu \in I^\perp$ and there exists $f \in RQ(S)$ with $f(x_j) = 1$ all j, $f(1) = f(0) = 0$. Finally $\delta(0) = 1-\delta(1)-f$ and so $\delta(0) \in RQ(S)$. □

1.10 Example: Let $S = \{1,s,0\} \cup \{x_j\}_{j\in\theta}$ (θ arbitrary) with the following rules of multiplication $1\cdot y = y$, $x_j^2 = s$ $(j \in \theta)$, all other products = 0. Then $c_c(S) \subset RQ(S)$. (This semigroup was invented by Macri; see Young [1].)

Proof. I_p is generated by the polynomials $z_j^2 - z_k^2$ $(j,k \in \theta)$, $z_iz_jz_k - z_iz_j$ $(i \neq j)$. The equivalence classes of Z_+^θ corresponding to the points of S are as follows $0 \in Z_+^\theta \leftrightarrow 1 \in S$, $e_j \in Z_+^\theta \leftrightarrow x_j \in S$, $s \in S$ correspond to $\{2e_j : j \in \theta\}$, $0 \in S$ corresponds to the complement in Z_+^θ of the previously defined classes. As in Example 1.9 we can show that $\delta(1)$, $\delta(x_j) \in RQ(S)$ $(j \in \theta)$, and that there exists $f \in RQ(S)$ with $f(x_j) = 1$ $(j \in \theta)$ and $f(1) = f(s) = f(0) = 0$. Thus it remains to show that $\delta(s) \in RQ(S)$. Indeed define $\nu \in M(\mathbb{T}^\theta)$ to be the Haar measure of the compact subgroup $G = \{(\lambda\varepsilon_j)_{j\in\theta} : \lambda \in \mathbb{T},\ \varepsilon_j = \pm 1\}$. Then $\hat{\nu} = 0$ or 1, and $\hat{\nu}(\alpha) = 1$ exactly when $\Sigma_j \alpha_j = 0$ and each α_j is even $(\alpha \in Z^\theta)$. Now define $\mu \in I^\perp \subset M(\mathbb{T}^\theta)$ by $d\mu = \bar{z}_{j_0}^2\, d\nu$ (for some $j_0 \in \theta$), then $\hat{\mu}(2e_j) = 1$ for all $j \in \theta$ and $\hat{\mu} = 0$ on $Z_+^\theta \setminus \{2e_j : j \in \theta\}$. Thus $\delta(s) \in RQ(S)$. □

We conjecture that $RQ(S)$ separates points for any discrete semigroup S.

There is a natural cone contained in $RQ(S)$, namely those elements achieving their norm at 1. We will use the letter P to denote this set, motivated by the following positive measure representation: Consider $RQ(S)$ as the dual of A_θ/I as in 1.4, and let $\pi : M(\mathbb{T}^\theta) \to A_\theta^*$ (dual of A_θ) be the canonical quotient map. Then $RQ(S) \cong I^\perp / A_\theta^\perp$ (considering $I^\perp$, $A_\theta^\perp \subset M(\mathbb{T}^\theta)$). Also $I^\perp = \pi^{-1}N$, where N is the annihilator of I in A_θ^*, and so $N \cong RQ(S)$. Suppose $f \in RQ(S)$ with $||f||_Q = f(1)$ and interpret f as an element of $N \subset A_\theta^*$. By the Hahn-Banach theorem there exists $\mu \in M(\mathbb{T}^\theta)$ such that $||\mu|| = ||f||_Q$ and $f = \pi\mu$. But $\int_{\mathbb{T}^\theta} 1\, d\mu = f(1) = ||\mu||$ so $\mu \geq 0$; in addition $\pi\mu \in N$ implies $\mu \in I^\perp \subset M(\mathbb{T}^\theta)$. Conversely if $\mu \in I^\perp \subset M(\mathbb{T}^\theta)$ and $\mu \geq 0$ then $\pi\mu \in RQ(S)$ (see 1.8) and $||\mu|| \geq ||\pi\mu||_Q \geq \pi\mu(1) = \int_{\mathbb{T}^\theta} 1\, d\mu = ||\mu||$ so $||\pi\mu||_Q = \pi\mu(1)$.

1.11 <u>Definition</u>: Let S be a semitopological semigroup with 1. Define $P(S) = \{f \in RQ(S) : f(1) = ||f||_Q\}$. It is clear that $P(S)$ is a norm-closed cone ($f, g \in P(S)$, $t > 0$ imply $f + tg \in P(S)$) and is closed under multiplication (observe $||f||_Q ||g||_Q = f(1)g(1) \leq ||fg||_Q \leq ||f||_Q ||g||_Q$ for $f,g \in P(S)$).

We show in Chapter 8 (8.1.11) that $P(S)$ (linearly) spans $RQ(S)$, as a consequence of a Hilbert space representation theorem for dual Q-algebras (8.1.8). The representation theorem for Q-algebras is attributed to Cole (see Bonsall and Duncan [1, p. 272]) and is given below.

1.12 Theorem: *Let A be a function algebra with 1 and let I be a closed ideal in A. Then there exists a Hilbert space $\mathcal{H}$ so that the Q-algebra A/I is isometrically isomorphic to a closed subalgebra of $\mathcal{B}(\mathcal{H})$ (the algebra of bounded linear operators on $\mathcal{H}$). Further the dual of A/I may be identified with the space of ultra-weak-operator (UWO) continuous linear functionals (restricted to the image of A/I in $\mathcal{B}(\mathcal{H})$).*

Proof. We first construct a representation of A/I on some Hilbert space and then prove the isometry.

Consider A as a subalgebra of $C(X)$ for some compact Hausdorff space X. Fix $\mu \in M_p(X)$ (a probability measure on X), and define $H^2(\mu)$ to be the closure of A in $L^2(\mu)$. Let $\mathcal{H}(\mu)$ be the orthogonal complement of I in $H^2(\mu)$, thus $H^2(\mu) = \mathcal{H}(\mu) \oplus \overline{I}$ (where $\overline{I}$ is the closure of I in $H^2(\mu)$). For each $f \in A$ define T_f to be the operator: $g \mapsto fg$ ($g \in H^2(\mu)$), then $||T_f|| \leq ||f||_\infty$. Define E to be the orthogonal projection $H^2(\mu) \to \mathcal{H}(\mu)$. For $f \in A$, let $\psi_f = ET_f|\mathcal{H}(\mu)$, then $\psi_f \in \mathcal{B}(\mathcal{H}(\mu))$ and $||\psi_f|| \leq ||f||_\infty$. Since I is an ideal we have that $T_f\overline{I} \subset \overline{I}$ ($f \in A$), and thus $\psi: f \mapsto \psi_f$ is a homomorphism of A into $\mathcal{B}(\mathcal{H}(\mu))$. Indeed, let $f,g \in A$, $h \in \mathcal{H}$, then

7.1.12

$\psi_{fg}\, h = E(fgh) = E(f(Egh) + f(1-E)gh) = E(f(Egh)) = \psi_f\psi_g h$ (since $(1-E)gh \in \bar{I}$). Further for $f \in I$, $\psi_f = 0$, since $T_f H^2(\mu) \subset \bar{I}$, thus I is in the kernel of ψ. For any $f \in A$ the bound $||\psi_f|| \leq ||f||_{A/I}$ holds.

Let K be the Hilbert space direct sum of the $H(\mu)$ ($\mu \in M_p(X)$), and let ϕ be the homomorphism of A/I into $B(K)$ given by ψ in each summand of K. We must now show that ϕ is isometric.

Fix some $f \in A$ with $||f||_{A/I} = 1$. By Hahn-Banach and Riesz theorems there exists $\nu \in M(X)$ with $||\nu|| = 1$, $\int_X f d\nu = 1$, and $\int_X g d\nu = 0$ for all $g \in I$. There exists a sequence $\{h_n\} \subset I$ such that $||f-h_n||_\infty < 1+1/n$. Let g be a weak-* cluster point of $f-h_n$ in $L^\infty(|\nu|)$ then $||g||_\infty \leq 1$. Also $\int_X g d\nu$ is a cluster point of $\int_X (f-h_n) d\nu = \int_X f d\nu = 1$, thus $\int_X g d\nu = 1$. But $||g||_\infty \leq 1$, $||\nu|| = 1$ and $\int_X g d\nu = 1$ imply that $g d\nu \geq 0$, $g d\nu = d|\nu|$, and $d\nu = \bar{g} d|\nu|$ (note $|g| = 1$ $|\nu|$ - a.e.). Let $\mu = |\nu| \in M_p(X)$. Note that since g is a weak-* cluster point of A in $L^\infty(\mu)$ it is also a weak cluster point of A in $L^2(\mu)$, but $H^2(\mu)$ is weakly closed hence $g \in H^2(\mu)$. Indeed $g \in H(\mu)$ since $\int_X h\bar{g} d\mu = \int_X h d\nu = 0$ ($h \in I$). Further f-g is a weak cluster point in $L^2(\mu)$ of $\{h_n\} \subset I$, thus $f-g \in \bar{I}$. Writing $f = g+(f-g)$ we see that $Ef = g$ and thus $||\psi_f 1||_2 = ||Ef||_2 = ||g||_2 = 1$. Thus $||\psi_f|| = 1$ and so ϕ is isometric (since $H(\mu)$ is one of the summands of K).

The second dual of $B(K)$ is isomorphically isometric to a subalgebra of $B(H)$ (1.2.7), some Hilbert space H, and the UWO-topology on $B(H)$ restricted to $B(K)$ is identical to the weak topology on $B(K)$. The above shows the dual of $B(K)$ maps onto $(A/I)^*$. Thus if we map A/I into $B(H)$ we obtain that the space of UWO-continuous linear functionals maps onto $(A/I)^*$. This means that given $\omega \in (A/I)^*$, there exist points $\{\eta_k\}_{k=1}^{\infty}$, $\{\xi_k\}_{k=1}^{\infty} \subset H$ such that $\Sigma_{k=1}^{\infty} ||\eta_k||\ ||\xi_k|| = ||\omega||$ and $\omega(f) = \Sigma_{k=1}^{\infty} \langle \phi_f \eta_k, \xi_k \rangle$ $(f \in A)$ (we use ϕ to denote the homomorphism $A/I \to B(H)$) (see Sakai [1, p. 39] or Chapter 8.§1). □

We now apply the theorem to discrete commutative semigroups, letting QA(S) be the Q-algebra in question.

1.13 Theorem: Let S be a discrete semigroup with 1, then there exists a Hilbert space H and a representation ϕ of S in the unit ball of $B(H)$ such that:

1) $||\Sigma_{x \in S} a_x \phi(x)|| = ||\Sigma_x a_x \delta(x)||_{QA}$ (for all $\Sigma_x a_x \delta(x) \in c_c(S)$);

2) for $\xi, \eta \in H$ the function $f: x \mapsto \langle \phi(x)\xi, \eta \rangle$ is in RQ(S) with $||f||_Q \leq ||\xi||\ ||\eta||$;

3) for each $f \in$ RQ(S) there exist sequences $\{\xi_n\}$, $\{\eta_n\} \subset H$ such that $\Sigma_{n=1}^{\infty} ||\xi_n||\ ||\eta_n|| = ||f||_Q$ and $f(x) = \Sigma_{n=1}^{\infty} \langle \phi(x)\xi_n, \eta_n \rangle$, $(x \in S)$;

4) the WO-closure of $\phi(S)$ is a compact semitopological semigroup isomorphic to the weak-* closure of S in RQ(S)* (see Theorem 6.2.10).

Note that condition 1) implies condition 2). We point out that there exists a Hilbert space representation ϕ of Z_+^n (some $n \geq 3$) for which $||\Sigma_x a_x \phi(x)|| \leq ||\Sigma_x a_x \delta(x)||_{QA}$ fails to hold. We will return to this in the next chapter. There we also prove a theorem (8.1.10) for semitopological semigroups analogous to 1.13.

§2. Semigroups of type $\mathcal{U}$

When S is a semitopological semigroup that has a dense inverse semigroup then the theory of Q-representations adds nothing new to the L^∞-representations, in the sense that $RQ(S) = \mathcal{R}(S)$. However the defining condition of $RQ(S)$ does give a new characterization of $\mathcal{R}(S)$. We will prove these assertions in this section.

Using the Hilbert space representation theorem of the preceding section we will show that $C(\hat{S}) \cong QA(S)$ for any discrete inverse semigroup S with 1. This means that $\hat{S}$ is a "bounded interpolation" set for A_θ. By way of illustration we will give a direct proof in the case of a finitely generated abelian group.

2.1 <u>Theorem: Let</u> S <u>be a discrete inverse semigroup with</u> 1, <u>then</u> $C(\hat{S}) \cong QA(S)$ <u>and</u> $RQ(S) = \mathcal{R}(S)$ <u>(with identical norms)</u>.

Proof. Let ϕ be the representation of S on the Hilbert space with the properties stated in Theorem 1.13. For any $x \in S$, $e = xx'$ is an idempotent, thus $\phi(e)$ is an idempotent in $\mathcal{B}(\mathcal{H})$ with $||\phi(e)|| \leq 1$, but this implies that $\phi(e)$ is an orthogonal projection (see the methods of Theorem 3.1.5). Further $\phi(x)\phi(e) = \phi(xe) = \phi(x) = \phi(e)\phi(x)$ and $\phi(x)\phi(x') = \phi(x')\phi(x) = \phi(e)$ imply that $\phi(x') = \phi(x)^*$. Thus each $\phi(x)$ is normal and $\phi(x)^* = \phi(x')$. Let $\Sigma_x a_x \delta(x) \in c_c(S)$, then $\Sigma_x a_x \phi(x)$ is normal and hence the spectral radius of $\Sigma_x a_x \phi(x)$ equals the operator norm. However the spectral radius of $\Sigma_x a_x \phi(x)$ is equal to $\sup_{\psi \in \hat{S}} |\Sigma_x a_x \psi(x)|$ (see 1.6). Together with property 1), Theorem 1.13, this implies that $||\Sigma_x a_x \delta(x)||_{QA} = \sup_{\psi \in \hat{S}} |\Sigma_x a_x \psi(x)|$. By the Stone-Weierstrass theorem the completion of $c_c(S)$ in the $\hat{S}$-sup norm is $C(\hat{S})$. Thus $QA(S)$ is isometrically isomorphic to $C(\hat{S})$. By Theorem 4.5, $RQ(S)$

is the dual of $C(\hat{S})$ so that $RQ(S) = R(S)$ and $||f||_Q = ||f||_R$ $(f \in RQ(S))$. (See Theorem 1.4.) □

We continue with the hypotheses of Theorem 2.1 and recall that $\hat{S}$ may be identified with a compact subset of $\overline{U}^\theta$ (where $\{x_j\}_{j\in\theta}$ generates S) (see 1.6). The theorem says that $\hat{S}$ is a bounded interpolation set for A_θ: that is, given $f \in C(\hat{S})$ and $\varepsilon > 0$ there exists $g \in A_\theta$ such that $g|\hat{S} = f$ and $||g||_\infty < ||f||_\infty + \varepsilon$. Indeed, given a polynomial (in z) p and $\varepsilon > 0$, then there exists $g \in I_p$ such that $||p+g||_\infty < \sup_{\hat{S}}|p| + \varepsilon$. From this one can deduce that the set of restrictions of A_θ to S is closed in $C(\hat{S})$ and proceed by a standard argument (see Rudin [3, p. 133]; extend convergent sequences to Cauchy sequences in A_θ).

We give a direct argument for Theorem 2.1 on a finitely generated abelian group G.

2.2 Example: Let G be a finitely generated abelian group, then $QA(G) \cong C(\hat{G})$.

Proof. We begin with a lemma: Suppose $p \in P_n$ (some integer n) such that $E = \{\lambda \in \overline{U}^n : p(\lambda) = 1\} \subset T^n$ and $|p(\lambda)| < 1$ for $\lambda \in \overline{U}^n \setminus E$, then $A_n/\overline{I}(1-p) \cong C(E)$ (where $\overline{I}(1-p)$ is the closed ideal generated by 1-p). Indeed for any $q \in P_n$, qp^m has the same restriction to E as g, and with $||qp^m||_\infty$ arbitrarily close to $\sup_E|q|$ (taking m large enough), and $q - qp^m = q(1-p^m) \in I(1-p)$. An argument of Stout shows that $A_n|E$ is closed in E, and contains the functions $1, z_j, \overline{z_j}$ $(1 \leq j \leq n)$ (see Rudin [3, p. 132] or Stout [13, p. 221]) and thus $A_n/\overline{I}(1-p) \cong C(E)$. Further $\overline{I}(1-p)$ $= \{f \in A_n : f(E) = 0\}$.

7.2.3

Given a finite number of such polynomials $p_1,\cdots,p_m \in P_n$, such that $p_k(\lambda) = 1$ for $\lambda \in E_k \subset T^n$ and $|p_k(\lambda)| < 1$ for $\lambda \in \bar{U}^n \setminus E_k$ $(1 \le k \le m)$, let $p_0 = p_1p_2\cdots p_m$, then $\bar{I}\,(1-p_0) = I$, the closed ideal generated by $\{1-p_k : 1 \le k \le m\}$, and $A_n/\bar{I}\,(1-p_0) \cong C(E)$ where $E = \bigcap_{k=1}^{m} E_k$. Since $1-p_k = 0$ on E we have that $1-p_k \in \bar{I}\,(1-p_0)$ and so $I \subset \bar{I}\,(1-p_0)$. Further $1-p_1p_2\cdots p_m = (1-p_1) + p_1(1-p_2) + p_1p_2(1-p_3) + \cdots + p_1\cdots p_{m-1}\,(1-p_m) \in I$. Thus $I = \bar{I}\,(1-p_0)$.

By the structure theorem, G is a direct sum of Z^m and finitely many Z(k) (integers mod k). As semigroup generators for G we choose $\chi_1,\cdots,\chi_{2m+r}$ with the relations $\chi_j\chi_{m+j} = 1$, $1 \le j \le m$ and $\chi_{2m+j}^{k_j} = 1$, $1 \le j \le r$, where the k_j are the orders of the finite cyclic subgroups. We are concerned with the closed ideal I in A_{2m+r} generated by $\{z_jz_{m+j}-1 : 1 \le j \le m\}$ and $\{z_{2m+j}^{k_j}-1 : 1 \le j \le r\}$. The polynomials $z_jz_{m+j}-1$, $z_{2m+j}^{k_j}-1$ can be written in the form $-2(1-(z_jz_{m+j}+1)/2)$ and $-2(1-(z_{2m+j}^{k_j}+1)/2)$ respectively.

The polynomials $(z_jz_{m+j}+1)/2$ and $(z_{2m+j}^{k_j}+1)/2$ satisfy the hypotheses of the above lemma, thus $A_{2m+r}/I \cong C(E)$, where E is the set of points in T^{2m+r} where all these polynomials take the value 1. But E corresponds to $\hat{G}$ as in Theorem 2.1, so $QA(G) \cong C(\hat{G})$. □

2.3 <u>Theorem</u>: <u>Let</u> S <u>be</u> <u>a</u> <u>semitopological</u> <u>semigroup</u> <u>with</u> 1 <u>of</u> <u>type</u> $\mathcal{U}$ (<u>a</u> <u>dense</u> <u>inverse</u> <u>semigroup</u>), <u>then</u> $\mathcal{R}(S) = RQ(S)$ <u>with</u> <u>identical</u> <u>norms</u>.

Proof. Let $f \in \mathcal{R}(S)$ then $f \in RQ(S)$ and $||f||_Q \le ||f||_{\mathcal{R}}$ by Corollary 6.2.6. Conversely let $g \in RQ(S)$, and let U be a dense inverse semigroup of S, taken with the discrete topology, and with $1 \in U$. Then $g|U \in RQ(U)$ with $||g|U||_Q = ||g||_Q$ (Theorem 6.2.15). By Theorem 2.1, $g|U \in R(U)$ with $||g|U||_R = ||g|U||_Q$. But now $g|U \in R(U) \cap (C^B(S)|U)$ thus by Theorem 3.2.11 $g \in R(S)$ with $||g||_R = ||g|U||_R = ||g||_Q$. $\square$

2.4 Corollary: Let S be a semitopological semigroup with 1 of type U, and let $f \in C^B(S)$. Then $f \in R(S)$ if and only if $|\Sigma_{\alpha \in Z_+^n} a_\alpha f(x^\alpha)| \leq K_f \sup_{\lambda \in T^n} |\Sigma_\alpha a_\alpha \lambda^\alpha|$ for each set $x_1, \cdots, x_n \in S$, polynomials $\Sigma_\alpha a_\alpha z^\alpha \in P_n$ and some constant $K_f < \infty$. Further $||f||_R$ is the least value of K_f for which the bound holds.

Two important examples of such semigroups are locally compact abelian groups and idempotent semigroups for which the above gives a new characterization of Fourier-Stieltjes transforms ($R(G) = M(\hat{G})\hat{\ }$) and functions of bounded variation (see 3.3.16) respectively. Theorem 2.3 and Corollary 2.4 are due to Dunkl. The LCA version was announced by him in [1].

The reader should note that we have mentioned two kinds of positivity for functions on a semigroup S of type U. There is the set $R_+(S)$ of positive-definite functions (Chapter 3, §1) and the set $P(S)$ (1.11). Fortunately they coincide.

2.5 Proposition: Suppose S is a semitopological semigroup with 1 of type U, then $P(S) = R_+(S)$.

Proof. If f is positive-definite on S then $f \in R(S)$ and $f(1) = ||f||_R$ (see 3.1.5). By 2.3 we have $f \in RQ(S)$ and $f(1) = ||f||_R = ||f||_Q$ so $f \in P(S)$.

Conversely let $f \in P(S)$, and let U be a dense inverse semigroup of S with $1 \in U$ and U considered discrete. Then $f|U \in P(U)$ (note $||f|U||_Q = ||f||_Q$, see 6.2.16). But $f|U \in R(U)$ (by 2.3) and hence corresponds to a unique measure $\mu \in M(\hat{U})$ (see 4.5.2) with $\int_{\hat{U}} 1 d\mu = f(1) = ||f|U||_R = ||\mu||$, thus $\mu \geq 0$ and f is positive-definite on U (4.5.4), hence on S (by Definition 3.1.2).□

§3. Z_+^θ and $\mathbb{R}_+^n$ and their Rees quotients

We show in this section that $RQ(Z_+^\theta) = R(Z_+^\theta) = M(\mathbb{T}^\theta)^\wedge|Z_+^\theta$ for any index set θ, and $RQ(\mathbb{R}_+^n) = R(\mathbb{R}_+^n) = M(\mathbb{R}^n)^\wedge|\mathbb{R}_+^n$, for $n = 1,2,3,\cdots$. This makes it possible to determine RQ for Rees quotients of Z_+^θ and $\mathbb{R}_+^n$, for example, the nil-thread.

3.1 <u>Theorem</u>: <u>Let θ be an index set, then $QA(Z_+^\theta) = A_\theta$ and $RQ(Z_+^\theta) = M(\mathbb{T}^\theta)^\wedge|Z_+^\theta$ (with the restriction norm)</u>.

Proof. It is clear from the definition of $QA(Z_+^\theta)$ that $QA(Z_+^\theta) = A_\theta$ (the corresponding ideal $I_P = \{0\}$). Thus $RQ(Z_+^\theta) = A_\theta^*$ (Theorem 1.4). For each $f \in RQ(Z_+^\theta)$, there exists $\mu \in M(\mathbb{T}^\theta)$ with $||\mu|| = ||f||_Q$ and $\hat{\mu}(\alpha) = \int_{T^\theta} z^\alpha d\mu = f(\alpha)$ $(\alpha \in Z_+^\theta)$. □

3.2 <u>Corollary</u>: $RQ(Z_+^\theta) = R(Z_+^\theta)$.

Proof. Observe that A_θ is the completion of $\ell^1(Z_+^\theta)$ in the spectral norm. Now apply the above theorem and Theorem 4.5.1.

We need a lemma of independent interest for the case of $\mathbb{R}_+^n$. Suppose $E \subset \mathbb{R}$ and $\{x_1,\cdots,x_k\} \subset \mathbb{R}^n$, then let E-span $\{x_1,\cdots,x_k\}$ $= \{\Sigma_{j=1}^k u_j x_j : u_j \in E\} \subset \mathbb{R}^n$. Let Q denote the field of rational numbers, and let $Q_+ = \mathbb{R}_+ \cap Q$.

3.3 <u>Lemma</u>: <u>Suppose</u> $\{x_1,\cdots,x_k\} \subset \mathbb{R}^n_+$, (k,n <u>positive integers</u>), <u>then there exist</u> $y_1,\cdots,y_m \in \mathbb{R}^n_+$ <u>such that</u> $\{y_1,\cdots,y_m\}$ <u>is linearly independent over</u> Q <u>and</u> $x_j \in Z_+$-span $\{y_1,\cdots,y_m\}$, $1 \leq j \leq k$. <u>That is, each finite subset of</u> $\mathbb{R}^n_+$ <u>is contained in some finitely generated free subsemigroup of</u> $\mathbb{R}^n_+$.

Proof. Assume first that n = 1. We proceed by induction. Suppose that $u_1,\cdots,u_s \in \mathbb{R}_+$, are linearly independent over Q, and $x_j \in Q_+$-span $\{u_1,\cdots,u_s\}$ for some $\ell, j \leq \ell < k$. If $\{u_1,\cdots,u_s,x_{\ell+1}\}$ is linearly independent over Q put $u_{s+1} = x_{\ell+1}$ (this step begins the construction if s = 0). Otherwise there exist unique $a_j \in Q$, $1 \leq j \leq s$ with $x_{\ell+1} = \Sigma^s_{j=1} a_j u_j$. If $a_j \geq 0$ for all j, go on to $x_{\ell+2}$ and repeat the above procedure. If some $a_j < 0$ then $x_{\ell+1} > 0$ and at least one $a_j > 0$ (since $u_1,\cdots,u_s > 0$). For convenience, renumber the u's so that $a_j > 0$ for $1 \leq j \leq t$, $a_j \leq 0$ for $t+1 \leq j \leq s$. Let $v = \Sigma^s_{j=t+1}(-a_j)u_j$ thus $v \in Q_+$-span $\{u_{t+1},\cdots,u_s\}$.

Let Ω denote $\{w \in \mathbb{R}^t_+ : \Sigma^t_{i=1} w_i = 1\}$ and define a continuous function f on Ω by $f(w) = \min\{a_i u_i - w_i v : 1 \leq i \leq t\}$. Observe that $Q^t_+ \cap \Omega$ is dense in Ω and that $\{w \in \Omega : f(w) > 0\}$ is relatively open and nonempty; indeed $w_i = a_i u_i/(\Sigma^t_{j=1} a_j u_j)$ $(1 \leq i \leq t)$ defines a point for which $f(w) > 0$ (note that $a_i u_i - w_i v = x_{\ell+1} a_i u_i/(\Sigma^t_{j=1} a_j u_j) > 0$). Thus there exists $b \in Q^t_+ \cap \Omega$ such that $f(b) > 0$.

Now set $u_i' = a_i u_i - b_i v$ for $1 \leq i \leq t$, and put $u_i' = u_i$ for $t+1 \leq i \leq s$. We assert that $u_j \in Q_+$-span $\{u_1',\cdots,u_s'\}$ for $1 \leq j \leq s$, indeed $u_i = (1/a_i)u_i' + (b_i/a_i)\Sigma^s_{j=t+1}(-a_j)u_j'$ $(1 \leq i \leq t)$. Thus $\{u_1',\cdots,u_s'\}$ is linearly independent over Q, and each $u_j' > 0$ by construction. Also $x_j \in Q_+$-span $\{u_1',\cdots,u_s'\}$ for $1 \leq j \leq \ell+1$. Indeed $x_{\ell+1} = \Sigma^t_{j=1} u_j'$, and for $1 \leq j \leq \ell$,

$x_j \in Q_+$-span $\{u_1,\cdots,u_s\} \subset Q_+$-span $\{u_1',\cdots,u_s'\}$.

Repeat the above construction until $\ell+1 = k$, and thus obtain $u_1,\cdots,u_m \in \mathbb{R}_+$ which are linearly independent over Q and such that $x_j \in Q_+$-span $\{u_1,\cdots,u_m\}$, $(1 \le j \le k)$. Now let $y_i = u_i/K$ $(1 \le i \le m)$ where K is a positive integer chosen so that $x_j \in Z_+$-span $\{y_1,\cdots,y_m\}$ $(1 \le j \le k)$.

For $n > 1$ apply the above construction to each coordinate separately, that is, find an appropriate basis for the first coordinates of $x_1,\cdots,x_k$, then for the second, and so on up to the nth coordinates. (We note that this process does not necessarily produce a minimal basis.) □

H. Rosenthal [1] characterized restrictions of Fourier-Stieltjes transforms to subset of $\mathbb{R}$, and R. Doss [1] proved the general version of the theorem, given below.

3.4 <u>Theorem</u>: <u>Let</u> G <u>be a locally compact abelian group, let</u> Λ <u>be a Borel subset of</u> G <u>and let</u> f <u>be a Borel function defined on</u> Λ <u>satisfying</u>: $|\Sigma_{i=1}^n a_i f(x_i)| \le K||\Sigma_{i=1}^n a_i \delta_{x_i}||_{sp}$, (<u>for</u> $a_i \in \mathbb{C}$, $x_i \in \Lambda$, $1 \le i \le n$, $n = 1,2,\cdots$). <u>Then there exists</u> $\mu \in M(\hat{G})$ <u>such that</u> $\hat{\mu} = f$ <u>almost everywhere</u> (<u>Haar measure</u>) <u>and</u> $||\mu|| \le K$.

We note that the spectral norm of $\Sigma a_i \delta_{x_i}$ equals the sup-norm of $\Sigma a_i \hat{x}_i$ over $\hat{G}$ (see 5.1.4).

It would be interesting to find a proof easier than that of Doss in the case where Λ is a subsemigroup of G. What is needed is to show that f defines a spectral-norm bounded linear functional on $L^1(\Lambda)$ (the functions in $L^1(G)$ which are 0 off Λ). One should be able to exploit the fact that $L^1(\Lambda)$ is an algebra if Λ is a subsemigroup.

3.5 Theorem: $RQ(\mathbb{R}_+^n) = R(\mathbb{R}_+^n) = M(\mathbb{R}^n)^\wedge|\mathbb{R}_+^n$, for $n = 1,2,\cdots$.

Proof. We already know that $RQ(\mathbb{R}_+^n) \supset R(\mathbb{R}_+^n)$ $= R(\mathbb{R}^n)|\mathbb{R}_+^n = M(\mathbb{R}^n)^\wedge|\mathbb{R}_+^n$ (see 6.2.6, and 5.1.5 respectively). Let $f \in RQ(\mathbb{R}_+^n)$. Let $x_1,\cdots,x_k \in \mathbb{R}_+^n$, then by Lemma 3.3 there exists a subsemigroup $S_o \subset \mathbb{R}_+^n$ such that $x_1,\cdots,x_k \in S_o$ and $S_o \cong Z_+^m$ for some integer m. Now $f|S_o \in RQ(S_o)$ (see 6.2.13) and by Corollary 3.2 $|\Sigma_{i=1}^k a_i f(x_i)| \leq ||f||_Q ||\Sigma_{i=1}^k a_i \delta_{x_i}||_{sp}$, $a_1,\cdots,a_k \in \mathbb{C})$. The bound holds for any finite subset of $\mathbb{R}_+^n$, thus by Doss' theorem (3.4) there exists $\mu \in M(\mathbb{R}^n)$ with $||\mu|| \leq ||f||_Q$ and $\hat{\mu} = f$ almost everywhere on $\mathbb{R}_+^n$. Since $\hat{\mu}$ and f are both continuous, in fact $f = \hat{\mu}|\mathbb{R}_+^n$. Thus $RQ(\mathbb{R}_+^n) \subset M(\mathbb{R}^n)^\wedge|\mathbb{R}_+^n$. □

Suppose J is a closed ideal in a semitopological semigroup S (with 1). Then the Rees quotient S/J is the semigroup consisting of the quotient space S with J identified as a point (thus J is the zero element) and with the inherited multiplication (see Hofmann and Mostert [1, p.25]). Corollary 6.2.16 shows that $RQ(S/J) = \{f \in RQ(S): f \text{ is constant on } J\}$.

3.6 Theorem: Let J be an ideal in Z_+^θ (some index set θ), then $RQ(Z_+^\theta/J)$ separates the points of Z_+^θ/J.

Proof. Indeed for any $\alpha \in Z_+^\theta \setminus J$, there exists $\mu \in M(\mathbb{T}^\theta)$ such that $\hat{\mu}(\alpha) = 1$, $\hat{\mu}(\beta) = 0$ for all $\beta \in Z^\theta$, $\beta \neq \alpha$. Proposition 1.8 now shows that $RQ(Z_+^\theta/J)$ separates points. □

3.7 Theorem: Let J be a closed ideal in $\mathbb{R}_+^n$ (n = 1,2,···), then $RQ(\mathbb{R}_+^n/J)$ separates the points of $\mathbb{R}_+^n/J$.

Proof. Let $x,y \in \mathbb{R}_+^n \setminus J$ with $x \neq y$. Then there exists a compact neighborhood V of x such that $V \subset \mathbb{R}_+^n \setminus J$ and $y \notin V$. Further there exists $\mu \in M(\mathbb{R}^n)$ such that $\hat{\mu}(x) = 1$, $0 \leq \hat{\mu} \leq 1$ and $\hat{\mu} = 0$ off V (indeed $\mu \in L^1(\mathbb{R}^n)$). Let $f = \hat{\mu}|\,\mathbb{R}_+^n$ then $f \in RQ(\mathbb{R}_+/J)$ and f separates x from y and J. □

3.8 Corollary: Suppose J is a closed ideal in $\mathbb{R}_+^n$ (n = 1,2,···) and $\mathbb{R}_+^n \setminus J$ is bounded (for example, if n = 1 and J = $[1,\infty)$, so that $\mathbb{R}/J$ is the nil-thread) then $\mathbb{R}_+^n/J$ is isomorphic to a compact subsemigroup of (the ball of) a dual Q-algebra.

Proof. Observe that $\mathbb{R}_+^n/J$ is compact and that $RQ(\mathbb{R}_+^n/J)$ separates points. Thus $\mathbb{R}_+^n/J$ has a faithful continuous Q-representation in $RQ(\mathbb{R}_+^n/J)^*$ (see Theorem 6.2.10). The compactness of $\mathbb{R}_+^n/J$ implies that the map is a homeomorphism. □

We consider the nil-thread, $\mathbb{R}_+/[1,\infty)$, in more detail. Let $\mu \in M(\mathbb{R})$ with $\hat{\mu}$ constant on $[1,\infty)$ with value c. Then $(\mu - c\delta_0)^\wedge$ is zero on $[1,\infty)$ hence is absolutely continuous that is, in $L^1(\mathbb{R})$ (see Rudin [2, p. 202]). Thus any function in $RQ(\mathbb{R}_+/[1,\infty))$ is a constant plus the restriction of $\hat{g}$ to $[0,1]$, where $g \in L^1(\mathbb{R})$ and $\hat{g} = 0$ on $[1,\infty)$.

Suppose that S is a compact semitopological semigroup with 1 such that RQ(S) is a regular algebra, that is, given $x \in S$, and a neighborhood V of x then there exists $f \in RQ(S)$ with $f(x) = 1$ and $f(y) = 0$ for $y \notin V$. For example, compact abelian groups, M = $[0,1]$ with $xy = \min(x,y)$, and the nil-thread $\mathbb{R}_1 = \mathbb{R}_+/[1,\infty)$

have this property. A Cartesian product of such semigroups will also have this property: indeed let $S = S_1 \times S_2$ and let $f \in RQ(S_1)$, then the function g on S defined by $g(x,y) = f(x)$, $(x \in S_1, y \in S_2)$ is in RQ(S) (the projection $S \to S_1$ is a homomorphism; apply 6.2.14). If RQ(S) is regular, and J is a closed ideal in S then RQ(S/J) separates the points of J. In particular RQ(D) separates the points of D, where D is the Rees quotient of $\mathbb{R}_1 \times M \times \mathbb{T}$ modulo $J = \{(x,y,z): x = 1 \text{ or } y = 0\}$ (note 1 is the zero element of $\mathbb{R}_1$). Thus if S is a uniquely divisible compact semigroup and satisfies certain additional assumptions (see D. Brown and M. Friedberg [1]) then RQ(S) separates the points of S, and S has a faithful representation in RQ(S)*.

§4. Further questions

4.1: If S is a discrete semigroup does RQ(S) separate the points of S? We note that if S is semitopological, then RQ(S) may fail to separate the points of S. Indeed this happens for S being the weakly almost periodic compactification of any noncompact locally compact abelian group (see Chapter 5, §2) and note $RQ(S) = R(S)$ (Theorem 2.3)).

4.2: If S_0 is a closed subsemigroup of a semitopological semigroup S does $RQ(S_0)$ extend to RQ(S), that is, does $RQ(S_0) = RQ(S)|S_0$?

4.3: Let S be a semitopological semigroup. Say a closed subset $E \subset S$ is an RQ-interpolation set (I-set for short) if $RQ(S)|E = C^B(E)$. For a given semigroup S are there any infinite I-sets ? This concept is the generalization of Sidon and Helson sets on locally compact abelian groups. It may happen that S is itself an interpolation set, see 1.9 and Dunkl [2].

4.4: For a given semigroup S, is RQ(S) a dual space ? For example, RQ(S) is the dual of QA(S) if S is discrete, and RQ(M) is not a dual space for M = [0,1] with the "min" operation.

4.5: If RQ(S) is a dual space, then the unit ball of RQ(S) has lots of extreme points. Some semicharacters are extreme points. It would be interesting to characterize extreme points for particular semigroups, especially those which are not separative. Are the extreme points a (multiplicative) subsemigroup of RQ(S) ?

Chapter 8. Hilbert space representations

Some time ago, von Neumann [1] showed that any contraction operator on a Hilbert space determined a representation of the disc algebra A_1. It was later shown that this is an easy consequence of the existence of a unitary dilation of a contraction. Naturally, the several-variable conjecture reared its challenging head, that is, given n commuting contractions, do they give a representation of the polydisc algebra A_n, or better yet, does there exist a joint dilation by commuting unitary operators? Eventually Parrott [1] found a counterexample to the latter question for n = 3, but his example did give a representation for A_3. Varopoulos [1] showed that the first part of the question fails for some $n \geq 3$, and then together with S. Kaijser showed by example it fails for n = 3.

Meanwhile, dilation theory was investigated from the point of view of operator-valued positive-definite functions on groups by Foiaş and Sz.-Nagy [1], and from the point of view of completely contractive and completely positive mappings by Arveson [1,2]. Interesting theorems resulted.

In this chapter we would like to give the reader a glimpse of these theories. In the context of semigroups, we classify representations with respect to various boundedness and dilation properties. These ideas will be connected to the theory of positive-definite functions (Chapter 3) and RQ(S) (Chapters 6,7). The hardest problems (in a sense, the only problems) concern the representations of subsemigroups of semigroups of type $\mathcal{U}$, which are not themselves type $\mathcal{U}$. The theory is essentially complete only for Z_+ and $\mathbb{R}_+$.

8.1.2

The chapter begins with basic facts about topologies on spaces of operators on Hilbert space.

In Section 1 we prove an existence theorem for certain Hilbert space representations having a close relationship with RQ(S). Roughly, given a commutative semitopological semigroup S with 1, there exists a weakly continuous Hilbert space representation of S isomorphic to the representation of S in RQ(S)* (see Theorem 6.2.10).

§1. General theory

1.1 Definition: i) Let H be a Hilbert space and let $B(H)$ be the algebra of bounded operators on H. Let $\langle\cdot,\cdot\rangle$ and $|\cdot|$ denote the inner product and norm on H, respectively. Thus $B(H)$ is normed by $||A|| = \sup\{|A\xi|:\xi \in H, |\xi| \leq 1\}$. Denote the unit ball $\{A \in B(H):||A|| \leq 1\}$ by $B_1(H)$.

ii) The weak operator (WO) topology on $B(H)$ is defined by: a net $\{A_\alpha\} \subset B(H)$ converges to A in WO if and only if $\langle A_\alpha\xi,\eta\rangle \overset{\alpha}{\to} \langle A\xi,\eta\rangle$, $(\xi,\eta \in H)$.

iii) The ultraweak operator (UWO) topology on $B(H)$ is defined by: $A_\alpha \overset{\alpha}{\to} A$ in UWO if and only if $\Sigma_{j=1}^{\infty}\langle A_\alpha\xi_j,\eta_j\rangle \overset{\alpha}{\to} \Sigma_{j=1}^{\infty}\langle A\xi_j,\eta_j\rangle$ for each pair of sequences $\{\xi_j\}$, $\{\eta_j\} \subset H$ with $\Sigma_{j=1}^{\infty}|\xi_j||\eta_j| < \infty$. Thus the UWO-topology is stronger than the WO-topology.

The following is well-known (Sakai [1, p. 38]).

1.2 Proposition: i) The WO and UWO topologies coincide on bounded sets in $B(H)$, in particular on $B_1(H)$.

ii) The WO topology is defined by the WO-continuous linear

functionals, and these are given by:

$$\omega(A) = \Sigma_{j=1}^{n} \langle A\xi_j, \eta_j \rangle,\ n = 1,2,\cdots;\ \xi_j, \eta_j \in H;$$

iii) *The UWO-topology is defined by the UWO-continuous linear functionals, and these are given by:*

$$\omega(A) = \Sigma_{j=1}^{\infty} \langle A\xi_j, \eta_j \rangle, \quad \text{where } \Sigma_{j=1}^{\infty} |\xi_j||\eta_j| < \infty.$$

iv) *Let* $B(H)_*$ *be the space of UWO-continuous linear functionals considered as a subspace of* $B(H)^*$, *the dual space of* $B(H)$, *then* $B(H)_*$ *is closed and its dual space is* $B(H)$. *Thus* $B_1(H)$ *is UWO (and WO) compact.*

v) *Multiplication in* $B(H)$ *is separately UWO and WO continuous Thus* $B_1(H)$, *with the WO-topology, is a compact noncommutative semitopological semigroup.*

vi) *The adjoint operation on* $B(H)$ *is WO and UWO continuous.*

Future references to $B_1(H)$ as a semigroup will implicitly mean the operation of multiplication and the WO (equivalently, UWO) topology. We will study $B_1(H)$ as an object for representations of commutative semitopological semigroups. Four increasingly restrictive properties for such representations will be defined, and examples illustrating mutual differences will be discussed. Throughout this chapter we will use the symbols S, S_0, S_1 to denote semitopological commutative semigroups with identity 1.

1.3 *Definition*: A *type* A *representation* of S is a continuous homomorphism ϕ of S into $B_1(H)$, with $\phi 1 = I$ (the identity in $B(H)$). This means: i) $\phi x \in B(H)$ with $||\phi x|| \leq 1$ $(x \in S)$;

ii) $\phi(xy) = (\phi x)(\phi y)$ $(x,y \in S)$; iii) $x_\alpha \overset{\alpha}{\to} x$ in S implies $\phi x_\alpha \overset{\alpha}{\to} \phi x$ (WO).

1.4 <u>Proposition</u>: <u>Let</u> ϕ <u>be a type</u> A <u>representation of</u> S <u>and</u> $\phi^\#$ <u>the bounded linear map</u> $\mathcal{B}(H)_* \to C^B(S)$ <u>given by</u> $\phi^\#\omega(x) = \omega(\phi x)$ $(\omega \in \mathcal{B}(H)_*, x \in S)$. <u>Then</u> $\phi^\#\mathcal{B}(H)_* \subset \mathrm{WAP}(S)$.

Proof. Fix $\omega \in \mathcal{B}(H)_*$. We note $|\phi^\#\omega(x)| = |\omega(\phi x)| \leq ||\omega|| \; ||\phi x|| \leq ||\omega||$ $(x \in S)$ so $\phi^\#\omega$ is bounded. It is continuous because $x_\alpha \overset{\alpha}{\to} x$ in S implies $\phi x_\alpha \overset{\alpha}{\to} \phi x$ (UWO). Let $E \subset \mathcal{B}(H)_*$ be the set $\{\omega_B : \omega_B(A) = \omega(AB), (A \in \mathcal{B}(H)), B \in \mathcal{B}_1(H)\}$. Since multiplication is separately UWO-continuous and $\mathcal{B}_1(H)$ is UWO-compact we see that E is weakly (i.e., $\sigma(\mathcal{B}(H)_*, \mathcal{B}(H))$ compact). The set of translates of $\phi^\#\omega$ equals $\{\phi^\#(\omega_B) : B = \phi y, y \in S\} \subset \phi^\# E$, but $\phi^\# E$ is weakly compact in $C^B(S)$, since $\phi^\#$ is weakly continuous (Dunford and Schwartz [1, p. 422]). Therefore $\phi^\#\omega \in \mathrm{WAP}(S)$. □

The above shows that the matrix entry functions of ϕ, that is functions of the form $x \mapsto \langle \phi x\xi, \eta \rangle$, are in WAP(S).

1.5 <u>Definition</u>: A <u>type</u> B <u>representation</u> of S is a type A representation which also satisfies:

$$(1.5^*) \qquad ||\Sigma_{\alpha \in Z_+^n} a_\alpha \phi(x^\alpha)|| \leq ||\Sigma_\alpha a_\alpha z^\alpha||_\infty$$

for each $n = 1,2,\cdots, x_1,\cdots,x_n \in S$, $\Sigma_\alpha a_\alpha z^\alpha \in P_n$ (notation of Chapter 6). (This property will also be referred to as being "Q-bounded".)

In the language of operator theory this definition says

$\{\phi x : x \in S\}$ is a commuting set of contractions for which the polydisc is a spectral set (this ignores the topology on S).

1.6 Proposition: Let ϕ be a type B representation of S, then $\phi^{\#}\mathcal{B}(H)_* \subset RQ(S)$ (see 1.4), and $||\phi^{\#}\omega||_Q \leq ||\omega||$ $(\omega \in \mathcal{B}(H)_*)$.

Proof. Let $\omega \in \mathcal{B}(H)_*$, then there exist sequences $\{\xi_j\}\{\eta_j\} \subset H$ with $\omega(A) = \Sigma_{j=1}^{\infty}\langle A\xi_j,\eta_j\rangle$ and $\Sigma_{j=1}^{\infty}|\xi_j||\eta_j| = ||\omega||$. It is clear that $\phi^{\#}\omega \in C^B(S)$. For any $n = 1,2,\cdots$, $x_1,\cdots,x_n \in S$, $\Sigma_\alpha a_\alpha z^\alpha \in P_n$ we have $|\Sigma_\alpha a_\alpha \phi^{\#}\omega(x^\alpha)|$
$= \Sigma_{j=1}^{\infty}\langle \Sigma_\alpha a_\alpha \phi(x^\alpha)\xi_j,\eta_j\rangle \leq \Sigma_{j=1}^{\infty}||\Sigma_\alpha a_\alpha\phi(x^\alpha)||\ |\xi_j||\eta_j|$
$\leq ||\Sigma_\alpha a_\alpha z^\alpha||_\infty ||\omega||$. □

Here are some conditions equivalent to (1.5*):

1.7 Theorem: Let ϕ be a type A representation of S, then the following are equivalent:

1) ϕ is type B;
2) (1.5*) holds for $x_1,\cdots,x_n \in S_0$, a dense subsemigroup of S;
3) (1.5*) holds for $x_1,\cdots,x_n \in \{x_j\}_{j\in\theta}$, a set of generators for S;
4) $||\Sigma_{x\in S}a_x\phi(x)|| \leq ||\Sigma_{x\in S}a_x\delta(x)||_{QA}$ (see 7.1.3). for each $\Sigma_x a_x\delta(x) \in c_c(S_d)$ (finite sum).

Proof. Clearly (1) implies (2) and (3). We first show (2) implies (1): Fix $\xi,\eta \in H$ and let $f(x) = \langle(\phi x)\xi,\eta\rangle$, then by hypothesis $f \in C^B(S)$ and $f|S_0 \in RQ(S_0)$ (by 1.6), with $||f|S_0||_Q \leq |\xi||\eta|$. By Theorem 6.2.15, $f \in RQ(S)$ with $||f||_Q \leq |\xi||\eta|$.

8.1.7

Let $n = 1,2,\cdots,x_1,\cdots,x_n \in S$, $\Sigma_\alpha a_\alpha z^\alpha \in P_n$. By the definition of RQ, $|<(\Sigma_\alpha a_\alpha \phi(x^\alpha))\xi,\eta>| = |\Sigma_\alpha a_\alpha f(x^\alpha)| \leq |\xi||\eta| \; ||\Sigma_\alpha a_\alpha z^\alpha||_\infty$. Since this holds for any $\xi,\eta \in H$ we conclude that $||\Sigma_\alpha a_\alpha \phi(x^\alpha)|| \leq ||\Sigma_\alpha a_\alpha z^\alpha||_\infty$, that is (1) holds.

Next we show (3) implies (4); Let $\{x_j\}_{j\in\theta}$ be a set of generators for S satisfying (3). As in Theorems 6.2.8 and 7.1.4 we have a homomorphism of Z_+^θ onto S_d (S taken discrete). The hypothesis shows that the homomorphism $\psi:P_\theta \to B(H)$ given by $\psi(\Sigma_\alpha a_\alpha z^\alpha) = \Sigma_\alpha a_\alpha \phi(x^\alpha)$ extends to a bounded homomorphism ψ' of $A_\theta \to B(H)$. The kernel of ψ' contains the kernel I_P of the canonical homomorphism $P_\theta \to c_c(S_d)$, thus ψ' is bounded in the quotient norm of A_θ/I (where I = closure I_P). This shows that ϕ is bounded in the QA-norm on $c_c(S_d)$ (see 7.1.3).

Finally we show (4) implies (1): This is almost obvious from 7.1.5. Sketchily, fix $\xi,\eta \in H$ then $f:x \mapsto <\phi x\xi,\eta>$ $(x \in S)$ is in RQ(S) and now apply the technique used above in (2) implies (1).□

A classical theorem of von Neumann [1] asserted that the unit disc is a spectral set for any contraction (i.e., element of $B_1(H)$). Combined with 1.7 (3) this says that any type A representation of Z_+ (or a quotient of Z_+) is also type B. The analogous theorem for Z_+^2 was proved by Andó [1]. It was a long-standing conjecture that the n-polydisc ($\overline{U}^n$) is a spectral set for n commuting contractions (a type A representation of Z_+^n) but N. Varopoulos [1] has shown that a counterexample exists. In particular, he showed that there exist: a finite-dimensional Hilbert space H, some $n \geq 3$, operators $A_1,A_2\cdots A_n \in B_1(H)$ such that the closed algebra generated by $I,A_1,\cdots,A_n$ is commutative but is not topologically and linearly isomorphic to a Q-algebra (later he and

Kaijser found an example for n = 3; see addendum to [1]). Thus there exists a type A representation of Z_+^n which is not type B.

Holbrook [1] showed that any type A representation of the "trivial" semigroup, our Example 7.1.9 (1x = x, xy = 0), is type B.

It would seem that type A non-B representations are very little understood and much remains to be done.

By extending Theorem 7.1.12 (Cole) to dual Q-algebras we can prove an existence theorem for type B representations; that is, there exists a type B representation ϕ so that $\phi^{\#}\mathcal{B}(H)_* = RQ(S)$. This was shown for discrete semigroups S in Theorem 7.1.13. First we give the extension of Cole's theorem.

1.8 Theorem: Let X be a compact Hausdorff space and let N be a closed C(X)-submodule of M(X) and suppose C(X) is the dual space of N. Denote the weak-* topology $\sigma(C(X),N)$ by τ, and suppose A is a τ-closed subalgebra of C(X) with $1 \in A$, and suppose I is a τ-closed ideal in A. Then A/I (a dual Q-algebra) is isometrically isomorphic to an UWO-closed subalgebra of $\mathcal{B}(H)$ for some Hilbert space H. Further the induced τ-topology on A/I (see 6.1.11) is isomorphic to the UWO-topology on $\mathcal{B}(H)$ restricted to the image of A/I.

Proof. We note that C(X) is a commutative W*-algebra (see 1.2.9), and multiplication is separately τ-continuous so A/I is a dual Q-algebra (see 6.1.16). Also N is an L-subspace of M(X), that is, $\mu \in N$, $\nu \in M(X)$ and $\nu \ll \mu$ implies $\nu \in N$ (because N is a closed C(X)-module). Thus N is the span of the probability measures in it, denoted by $N_p = M_p(X) \cap N$.

8.1.8

Fix $\mu \in N_p$ and define $H^2(\mu)$ to be the closure of A in $L^2(\mu)$. Let $\mathcal{H}(\mu)$ be the orthogonal complement of I in $H^2(\mu)$, thus $H^2(\mu) = \mathcal{H}(\mu) \oplus \bar{I}$ (where $\bar{I}$ is the closure of I in $H^2(\mu)$). For each $f \in A$ define T_f to be the operator: $g \mapsto fg$ $(g \in H^2(\mu))$, then $||T_f|| \leq ||f||_\infty$. We claim this homomorphism is τ-UWO continuous. Let $\omega \in \mathcal{B}(H^2(\mu))_*$ thus there exist sequences $\{g_j\}$, $\{h_j\} \subset H^2(\mu)$ with $\Sigma_{j=1}^\infty ||g_j||_2 ||h_j||_2 < \infty$ such that $\omega(T_f) = \Sigma_{j=1}^\infty \int_X f g_j \bar{h}_j d\mu$ $= \int_X f(\Sigma_j g_j \bar{h}_j) d\mu$. But $\Sigma_j g_j \bar{h}_j \in L^1(\mu) \subset N$, thus $f \mapsto \omega(T_f)$ is a τ-continuous linear functional $(f \in A)$.

Define E to be the orthogonal projection: $H^2(\mu) \to \mathcal{H}(\mu)$. For $f \in A$, let $\psi_f = ET_f|\mathcal{H}(\mu)$, then $\psi_f \in \mathcal{B}(\mathcal{H}(\mu))$ and $||\psi_f|| \leq ||f||_\infty$. Since I is an ideal we have that $T_f\bar{I} \subset \bar{I}$ $(f \in A)$, and thus $\psi: f \mapsto \psi_f$ is a homomorphism of A into $\mathcal{B}(\mathcal{H}(\mu))$. Indeed, let $f, g \in A$, $h \in \mathcal{H}(\mu)$ then $\psi_{fg}(h) = E(fgh) = E(f(E(gh)+(1-E)(gh)))$ $= E(fE(gh)) = \psi_f\psi_g h$, since $(1-E)(gh) \in \bar{I}$. Further, for $f \in I$, $\psi_f = 0$ since $T_f H^2(\mu) \subset \bar{I}$, thus I is contained in the kernel of ψ. For any $f \in A$, the bound $||\psi_f|| \leq ||f||_{A/I}$ holds.

Let $\mathcal{H} = \Sigma \oplus_{\mu \in N_p} \mathcal{H}(\mu)$, and let V be the homomorphism of A/I into $\mathcal{B}(\mathcal{H})$ given by ψ in each summand of $\mathcal{H}$. Then V is τ-UWO continuous, since a UWO-continuous linear functional on $\mathcal{B}(\mathcal{H})$ restricted to V(A/I) is an absolutely convergent sum of such functionals on individual $\mathcal{B}(\mathcal{H}(\mu))$'s, and these correspond to τ-continuous functionals on A (see above). We now show that V is isometric.

Fix some $f \in A$ with $||f||_{A/I} = 1$ and let $\varepsilon > 0$. Observe that C(X)/I is the dual space of $I^\perp \subset N$, thus there exists $\nu \in N$ with $||\nu|| = 1$, $1 \geq \int_X f d\nu > 1-\varepsilon^2$, and $\int_X g d\nu = 0$ for all $g \in I$. Choose

$h \in I$ so that $||f-h||_\infty < (1+2\varepsilon^2)^{1/2}$. By the Radon-Nikodym theorem, write $d\nu = \bar{k}d\mu$, where k is a Borel function on X, $|k| = 1$ and $\mu = |\nu| \in N_p$. Let $g = f-h \in A$, then $\int_X g d\nu = \int_X f d\nu > 1-\varepsilon^2$ and $\int_X |g-k|^2 d\mu = \int_X |1-\bar{k}g|^2 d\mu = \int_X (1-2 \text{ Re}(\bar{k}g)+|g|^2) d\mu$

$\leq 1-2 \text{ Re}\int_X g\bar{k}d\mu + 1 + 2\varepsilon^2 < 2+2\varepsilon^2-2+2\varepsilon^2 = 4\varepsilon^2$. Let E_1 be the orthogonal projection of $L^2(\mu)$ onto the orthogonal complement of I in $L^2(\mu)$, so that $E = E_1|H^2(\mu)$, and $E_1 k = k$ since $\int_X j\bar{k}d\mu = \int_X j d\nu = 0$ for all $j \in I$.

We claim $||Ef||_2 > 1-2\varepsilon$. Indeed $f = g+h$ so $Ef = Eg$ (since $h \in I$) and $||Ef-k||_2 = ||Eg-k||_2 = ||E_1 g - E_1 k||_2 \leq ||g-k||_2 < 2\varepsilon$; but $||k||_2 = 1$ so $||Ef||_2 > 1-2\varepsilon$.

Also $\psi_f(E1) = E(T_f E1) = E(T_f 1) = Ef$ thus $||\psi_f|| \geq ||Ef||_2/||E1||_2 \geq ||Ef||_2 > 1-2\varepsilon$. But $\varepsilon > 0$ was arbitrary, so $||Vf|| = 1 = ||f||_{A/I}$, thus V is an isometry of A/I into $\mathcal{B}(H)$.

We have already shown that for each $\omega \in \mathcal{B}(H)_*$ there exists $\nu \in N$ so that $\omega(Vf) = \int_X f d\nu$ $(f \in A)$, and $||\nu|| \leq ||\omega||$. This element ν is determined up to an element of $A^\perp \subset N$, and $\nu \in I^\perp \subset N$. Thus there is a bounded linear map $V_1 : \mathcal{B}(H)_* \to I^\perp/A^\perp$, with $||V_1|| \leq 1$. We identify $I^\perp/A^\perp$ with $(A/I)_*$, the predual of A/I, and observe that $V = V_1{}^*$ (the adjoint). Further V is one-to-one and has closed range, being an isometry, hence V_1 maps $\mathcal{B}(H)_*$ onto $(A/I)_*$ (theorem of Banach, see Dunford and Schwartz [1, p. 488]). Indeed V_1 is a quotient map, that is, V_1 induces an isometry of $\mathcal{B}(H)_*/\ker V_1$ onto $(A/I)_*$; observe V maps A/I isometrically onto $(\ker V_1)^\perp \subset \mathcal{B}(H)$ (a consequence of the closed range theorem) thus $V(A/I)$ is UWO-closed, and if $\omega \in \mathcal{B}(H)_*$ then the quotient norm $||\omega + \ker V_1|| = \sup\{|\omega(T)| : T \in (\ker V_1)^\perp$,

$||T|| \leq 1\} = \sup\{|\omega(Vf)|:f \in A/I,\ ||f|| \leq 1\}$
$= \sup\{|V_1\omega(f)|:f \in A/I,\ ||f|| \leq 1\} = ||V_1\omega||$. This shows that the τ-topology on A/I is isomorphic to the UWO-topology on V(A/I). □

1.9 Corollary: In the notation of Theorem 1.8 and its proof, $I^{\perp} \subset N$ is the linear span of the probability measures in it, namely $I^{\perp} \cap N_p$, and $A^{\perp}$.

Proof. It suffices to find for each $\mu \in I^{\perp} \subset N$, measures $\mu_j \in I^{\perp} \cap N_p$, numbers $c_j \geq 0$ $(j = 0,1,2,3)$ such that $\int_X f d\mu = \Sigma_{j=0}^{3} i^j c_j \int_X f d\mu_j$, $(f \in A)$. Fix $\mu \in I^{\perp} \subset N$, so by Theorem 1.8 there exists $\omega \in B(H)_*$ so that $\mu + A^{\perp} = V_1\omega$ (as elements of $I^{\perp}/A^{\perp}$). There exist sequences $\{\xi_k\}$, $\{\eta_k\} \subset H$ with $\Sigma_k|\xi_k||\eta_k| < \infty$ and $\omega(T) = \Sigma_k\langle T\xi_k,\eta_k\rangle$, $(T \in B(H))$. Further we assume $|\xi_k| = |\eta_k|$ for all k.

Recall $H = \Sigma \oplus_{\nu \in N_p} H(\nu)$, so there exists a sequence $\{\nu_m\} \subset N_p$ with $\xi_k, \eta_k \in \Sigma \oplus_{m=1}^{\infty} H(\nu_m)$ all k. Write

$\xi_k = (g_{km})_{m=1}^{\infty}$, $\eta_k = (h_{km})_{m=1}^{\infty}$ with g_{km}, h_{km} $H(\nu_m)$, (all m) and

$|\xi_k|^2 = \Sigma_m||g_{km}||_2^2$, $|\eta_k|^2 = \Sigma_m||h_{km}||_2^2$ (all k). For $f \in A$, we have $\omega(Vf) = \Sigma_k\langle Vf\xi_k,\eta_k\rangle = \Sigma_k\langle\Sigma_m fg_{km},\Sigma_m h_{km}\rangle = \Sigma_k\Sigma_m\int_X fg_{km}\bar{h}_{km}d\nu_m$

$= \Sigma_k\Sigma_m \frac{1}{4}\Sigma_{j=0}^{3} i^j\int_X f|g_{km}+i^j h_{km}|^2 d\nu_m$ (by the polar identity). We assert that $\Sigma_k\Sigma_m|g_{km}+i^j h_{km}|^2 d\nu_m$ converges absolutely in N each j; indeed $\Sigma_k\Sigma_m||\ |g_{km}+i^j h_{km}|^2 d\nu_m|| = \Sigma_k\Sigma_m||g_{km}+i^j h_{km}||_2^2$

$\leq \Sigma_k\Sigma_m(||g_{km}||_2+||h_{km}||_2)^2 \leq \{(\Sigma_k\Sigma_m||g_{km}||_2^2)^{1/2}+(\Sigma_k\Sigma_m||h_{km}||_2^2)^{1/2}\}^2$

(Cauchy-Schwartz inequality)

$= \{(\Sigma_k|\xi_k|^2)^{1/2}+(\Sigma_k|\eta_k|^2)^{1/2}\}^2 = 4\Sigma_k|\xi_k||\eta_k| < \infty$ (since $|\xi_k| = |\eta_k|$). We see that $d\lambda_j = \frac{1}{4}\Sigma_k\Sigma_m|g_{km}+i^j h_{km}|^2 d\nu_m$ is a positive measure in N, and annihilates I $(0 \leq j \leq 3)$. So we can write each $\lambda_j = c_j\mu_j$ with $c_j \geq 0$ and $\mu_j \in N_p \cap I^\perp$ (assuming I is proper in A). By construction $\int_X f d\mu = \omega(Vf) = \Sigma_{j=0}^3 i^j c_j \int_X f d\mu_j$, $(f \in A)$. □

The corollary was proved in another way (not using Hilbert space representations) by Seever [1]. We now apply the theorem to RQ(S).

1.10 <u>Theorem</u>: <u>Let</u> S <u>be a commutative semitopological semigroup with</u> 1. <u>Then there exists a type</u> B <u>representation</u> ϕ <u>of</u> S <u>on a Hilbert space</u> H <u>such that</u> $\phi^\# B(H)_* = RQ(S)$ <u>and</u> $\phi^\#$ <u>is a quotient map, that is, given</u> $f \in RQ(S)$, $\varepsilon > 0$ <u>there exist sequences</u> $\{\xi_j\},\{\eta_j\} \subset H$ <u>such that</u> $f(x) = \Sigma_{j=1}^\infty \langle(\phi x)\xi_j,\eta_j\rangle$ $(x \in S)$ <u>and</u> $\Sigma_j|\xi_j||\eta_j| < ||f||_Q + \varepsilon$.

Proof. Apply Theorem 1.8 to obtain a weak-*-UWO continuous representation V of RQ(S)* (a dual Q-algebra, Theorem 6.2.10) in $B(H)$. Recall there is a canonical homomorphism $\sigma:S \to RQ(S)^*$ (where $\sigma x(f) = f(x)$, $f \in RQ(S)$, $x \in S$), so define the homomorphism $\phi:S \to B(H)$ to be $V\circ\sigma$. Thus ϕ maps S continuously into $B_1(H)$. Theorem 1.8 shows V_1 maps $B(H)_*$ onto RQ(S), that is, given $f \in RQ(S)$, $\varepsilon > 0$, there exists $\omega \in B(H)_*$, $||\omega|| < ||f||_Q + \varepsilon$ such that $\chi(f) = \omega(V\chi)$ $(\chi \in RQ(S)^*)$, in particular, $f(x) = \sigma x(f) = \omega(V\sigma x) = \omega(\phi x)$ $(x \in S)$, so that $f = \phi^\#\omega$.

It is easy to see that ϕ satisfies (1.5*), indeed choose $x_1,\cdots,x_n \in S$, $\Sigma_\alpha a_\alpha z^\alpha \in P_n$ (some $n = 1,2,\cdots$), $\xi,\eta \in H$, then

$|\Sigma_{\alpha \in Z_+^n} a_\alpha \langle \phi(x^\alpha)\xi,\eta\rangle| = |\Sigma_\alpha a_\alpha f(x^\alpha)| \le ||f||_Q ||\Sigma_\alpha a_\alpha z^\alpha||_\infty$, where $f = \phi^\# \omega \in RQ(S)$ and $||f||_Q \le ||\omega|| = |\xi||\eta|$ ($\omega \in B(H)_*$ defined by $\omega(T) = \langle T\xi,\eta\rangle$, $T \in B(H)$). Thus ϕ is a type B representation. □

Recall $RQ(S)$ contains a "positive" cone $P(S) = \{f \in RQ(S) : f(1) = ||f||_Q\}$ (see 7.1.11). We can now show that $P(S)$ spans $RQ(S)$ (analogously to Corollary 1.9).

1.11 <u>Corollary</u>: <u>The linear span of</u> $P(S)$ <u>is all of</u> $RQ(S)$.

Proof. Consider the representation of S constructed in 1.10, and let $f \in RQ(S)$. There exists $\omega \in B(H)_*$ with $f = \phi^\#\omega$, and so there exist sequences $\{\xi_k\}\{\eta_k\} \subset H$ with $\Sigma|\xi_k||\eta_k| < \infty$ and $|\xi_k| = |\eta_k|$, all k, such that $\omega(T) = \Sigma_k \langle T\xi_k,\eta_k\rangle$. Define $\omega_j \in B(H)_*$, by $\omega_j(T) = \Sigma_k \langle T(\xi_k + i^j\eta_k),(\xi_k + i^j\eta_k)\rangle$, $(0 \le j \le 3)$, then $\omega = \frac{1}{4}\Sigma_{j=0}^3 i^j\omega_j$ (the sums converge absolutely, see 1.9). Let $f_j = \phi^\#\omega_j$ $(0 \le j \le 3)$ then $||f_j||_Q \ge f_j(1) = \Sigma_k|\xi_k + i^j\eta_k|^2 \ge ||\omega_j|| \ge ||f_j||_Q$, thus $f_j \in P(S)$, and $f = \frac{1}{4}\Sigma_{j=0}^3 i^j f_j$. □

1.12 <u>Theorem</u>: <u>Let</u> ϕ <u>be a type</u> A <u>representation of</u> S <u>such that</u> ϕx <u>is normal for each</u> x <u>in some dense subset of</u> S. <u>Then</u> ϕx <u>is normal for each</u> $x \in S$, ϕ <u>is type</u> B, <u>and</u> $\phi^\# B(H)_* \subset R(S)$.

Proof. Let ϕx be normal for each $x \in E$, where E is a dense subset of S. For any $y \in S$, $x \in E$ we have $(\phi x)^*(\phi y) = (\phi y)(\phi x)^*$ since ϕx is normal, by the theorem of Fuglede-Putnam-Rosenblum (Rudin [2, p. 300]). Given $x \in S$, let $\{x_\alpha\}$ be a net in E with $x_\alpha \xrightarrow{\alpha} x$, then $(\phi x_\alpha)^* \xrightarrow{\alpha} (\phi x)^*$, and so $(\phi x)^*(\phi x) = \lim_\alpha(\phi x_\alpha)^*(\phi x) = \lim_\alpha(\phi x)(\phi x_\alpha)^* = (\phi x)(\phi x)^*$. Thus ϕx is normal for each $x \in S$, and ϕx commutes with each $(\phi y)^*$ $(y \in S)$. Let A be the norm-closed subalgebra of $B(H)$ generated by $\{\phi x, (\phi x)^*: x \in S\}$, then A is a commutative C*-algebra. Let Ω be the (compact) maximal ideal space of A, and let ^ denote the Gelfand transform, an isometry of A onto $C(\Omega)$. The spectral theorem (Appendix B) asserts that for each $\xi,\eta \in H$, there exists a Borel measure $E_{\xi,\eta}$ on Ω with $||E_{\xi,\eta}|| \leq |\xi||\eta|$ such that $\langle T\xi,\eta\rangle = \int_\Omega \hat{T} dE_{\xi,\eta}$ for each $T \in A$. Fix $\xi,\eta \in H$. Observe that the function $x \mapsto \langle (T)(\phi x)\xi,\eta\rangle = \langle(\phi x)\xi, T^*\eta\rangle$ is continuous on S for each $T \in A$ (since ϕ is WO-continuous). Thus $x \mapsto \int_\Omega \hat{T}(\phi x)\hat{}\, dE_{\xi,\eta}$ is continuous on S for each $T \in A$. Let $\mu = c|E_{\xi,\eta}| \in M_+(\Omega)$ with c chosen so that $\mu \in M_p(\Omega)$. Then the set of $g \in L^1(\mu)$ such that $x \mapsto \int_\Omega (\phi x)\hat{}\, g dE_{\xi,\eta}$ is continuous contains $\hat{A} = C(\Omega)$ and is L^1-norm closed hence equals $L^1(\mu)$. Thus $(\hat{\phi},\mu,\Omega)$ is a L^∞-representation of S and so $f_{\xi,\eta}: x \mapsto \langle(\phi x)\xi,\eta\rangle = \int_\Omega (\phi x)\hat{}\, dE_{\xi,\eta}$ is in $R(S)$ with $||f_{\xi,\eta}|| \leq |\xi||\eta|$. Thus $f_{\xi,\eta} \in RQ(S)$ (see 6.2.6) with $||f_{\xi,\eta}||_Q \leq |\xi||\eta|$. Since this holds for any $\xi,\eta \in H$, we have that ϕ satisfies (1.5*) (as in proof of 1.7) and so ϕ is type B. Further any $\omega \in B(H)_*$ is of the form $\omega(T) = \Sigma_{j=1}^\infty \langle T\xi_j,\eta_j\rangle$ with $\Sigma_{j=1}^\infty |\xi_j||\eta_j| = ||\omega||$, so $\phi^\#\omega(x) = \Sigma_{j=1}^\infty f_{\xi_j,\eta_j}(x)$ $(x \in S)$ with absolute convergence in the R-norm; indeed $||\phi^\#\omega||_R \leq \Sigma_{j=1}^\infty ||f_{\xi_j,\eta_j}||_R \leq \Sigma_{j=1}^\infty |\xi_j||\eta_j| = ||\omega||$. □

The above theorem always applies to semigroups of type U (a dense inverse semigroup), since $\phi x' = (\phi x)^*$ for elements x of an inverse semigroup (see 3.1.5). On the other hand, it would not be generally useful for nonseparative semigroups.

§2. Dilation theory

If S is not separative then $R(S)$ does not separate S, so one has to consider representations involving nonnormal operators. However it may be possible to realize a set of nonnormal operators as projections of normal operators.

2.1 <u>Definition</u>: Let ϕ be a type A representation of S on H. Say ϕ is of <u>type</u> C if there exists a semigroup S_0, a type A representation ψ of S_0 by normal operators on some Hilbert space K containing H as a closed subspace (same norm), and a continuous homomorphism ρ of S_0 onto S (and $\rho 1 = 1$) such that the diagram

$$\begin{array}{ccc} S_0 & \xrightarrow{\rho} & S \\ \psi \downarrow & & \downarrow \phi \\ B_1(K) & \xrightarrow{\pi} & B_1(H) \end{array}$$

commutes (where π is the canonical map $B(K) \to B(H)$ given by $(\pi T)(\xi) = \mathrm{pr}(T\xi)$ $(\xi \in H)$, pr is the projection $K \to H$).

2.2 <u>Theorem</u>: <u>Each type</u> C <u>representation is also type</u> B.

Proof. Let the hypotheses and notation be as in 2.1. By 1.8, ψ is a type B representation of S_0. Let $y \in S_0$, $\xi,\eta \in H$ then by hypothesis $\langle(\psi y)\xi,\eta\rangle = \langle\phi(\rho y)\xi,\eta\rangle$. For some $n = 1,2,\cdots$ let $x_1,\cdots,x_n \in S$, and let $\Sigma_{\alpha \in Z_+^n} a_\alpha z^\alpha \in P_n$. Then there exist $y_1,\cdots,y_n \in S_0$ such that $\phi y_j = x_j$ $(1 \le j \le n)$. Thus

$|\Sigma_\alpha a_\alpha \langle\phi x^\alpha \xi,\eta\rangle| = |\Sigma a_\alpha \langle\psi y^\alpha \xi,\eta\rangle| = |\langle\Sigma_\alpha a_\alpha \psi y^\alpha \xi,\eta\rangle| \le ||\Sigma_\alpha a_\alpha z^\alpha||_\infty |\xi||\eta|$ for each $\xi,\eta \in H$. But then $||\Sigma_\alpha a_\alpha(\phi x^\alpha)|| \le ||\Sigma_\alpha a_\alpha z^\alpha||_\infty$ and thus ϕ is type B. □

8.2.2

For example, if S is discrete then the representation of S associated with QA(S) constructed in 7.1.11 and 7.1.12 is of type C. Indeed Z_+^θ plays the part of S_0, and $\Sigma \oplus \{L^2(\mu) : \mu \in M_p(\bar{T}^\theta)\}$ is the bigger Hilbert space containing $\Sigma \oplus H(\mu)$ on which S is represented.

Recall the representation of $\mathbb{R}_+/[1,\infty)$ (the nil-thread) constructed in 2.2.4. Let K be the Hilbert space $L^2(\mathbb{R})$ (Lebesgue measure), let H be the closed subspace $L^2((0,1))$, then the projection $pr: K \to H$ is nothing but truncation, $pr\, f(x) = f(x)$ for $x \in (0,1)$, $pr\, f(x) = 0$ for $x \notin (0,1)$. Let ψ be the representation of $\mathbb{R}_+$ on K given by translation: $(\psi t) f(x) = f(x+t)$ $(x \in \mathbb{R},\ t \in \mathbb{R}_+,\ f \in L^2(\mathbb{R}))$. Each ψt is unitary, thus normal. Define $\phi: \mathbb{R}_+ \to B(H)$ by $(\phi t) f = pr(\psi t) f$ $(f \in L^2((0,1)),\ t \in \mathbb{R}_+)$. It is clear that ϕ annihilates the ideal $[1,\infty)$ in $\mathbb{R}_+$ so can be taken to be a continuous representation of $\mathbb{R}_+/[1,\infty)$; indeed it agrees with T_t of Example 2.2.4. The definition of ϕ shows that it is of type C.

We need the concepts of completely positive linear maps of C*-algebras and completely contractive linear maps of their subspaces to be able to use the dilation theory of Arveson [1,2].

Let B be a C*-algebra with unit and let $B^{(m)}$ denote the algebra of $m \times m$ matrices with B entries. Then $B^{(m)}$ is again a C*-algebra. For example, if $B = B(H)$ (some Hilbert space H) then $B^{(m)}$ is realized as $B(H^m)$, where $H^m = \Sigma_{j=1}^m \oplus H$. If $B = C(X)$, (X a compact Hausdorff space) then $B^{(m)}$ is the algebra of continuous $m \times m$ complex matrix functions on X, with norm $||(f_{rs})||_\infty = \sup_{x \in X} ||(f_{rs}(x))||_{op}$ (where $||\cdot||_{op}$ is the ℓ^2-operator norm for $m \times m$ matrices). Suppose ϕ is a linear map of $B \to B'$

(B,B' are C*-algebras with units) then $\phi^{(m)}$ is the induced linear map $B^{(m)} \to B'^{(m)}$ (apply ϕ to each matrix entry). Say ϕ is completely positive if $\phi^{(m)} \geq 0$ for each $m = 1,2,3,\cdots$ (terminology of Stinespring [1]). It also makes sense to define completely positive maps on self-adjoint subspaces of B. Suppose ϕ is a linear map of a subspace $Y \subset B$ into B' then say that ϕ is contractive if $||\phi|| \leq 1$, completely contractive if $||\phi^{(m)}|| \leq 1$ for each $m = 1,2,3\cdots$ (where $\phi^{(m)}:Y^{(m)} \to B'^{(m)}$).

Stinespring [1] characterized the completely positive maps of C*-algebras into $B(H)$:

2.3 Theorem: *Let B be a C*-algebra with unit and let ϕ be a linear map of $B \to B(H)$. Then ϕ is completely positive if and only if there exists a *-representation ψ of B in $B(K)$, some Hilbert space K, and a bounded linear map $V:H \to K$ such that $\phi(f) = V^*\psi(f)V$, $(f \in B)$.*

Stinespring also showed:

2.4 Theorem: *Let ϕ be a positive linear map of $C(X) \to B(H)$ then ϕ is completely positive (X compact Hausdorff space).*

The main extension theorem of Arveson [1, p. 149] is:

2.5 Theorem: *Let Y be a closed self-adjoint linear subspace of B, a C*-algebra with unit 1, and $1 \in Y$. Then every completely positive linear map $\phi:Y \to B(H)$ has a completely positive linear extension $\phi_1:B \to B(H)$.*

2.6 *Corollary*: *Let* Y *be a linear subspace of* $\mathcal{B}$ *with* $1 \in Y$. *Then every completely contractive linear map* $\phi: Y \to \mathcal{B}(H)$ *for which* $\phi(1) = I$ *has a completely positive linear extension to* $\mathcal{B}$.

A sort of converse to 2.5 is (Arveson [1, p. 153]).

2.7 *Proposition*: *Let* ϕ *be a completely positive map of* $\mathcal{B} \to \mathcal{B}'$ ($\mathcal{B}, \mathcal{B}'$ *are* C*-*algebras with units*), *then* $||\phi^{(m)}|| = ||\phi(1)||$, *for each* $m = 1,2,\cdots$.

If, for example, $\phi(1) = 1$ in 2.7 then ϕ is completely contractive. If the range C*-algebra is commutative then more information is available. In fact, Arveson [1, p. 154] showed:

2.8 *Proposition*: *Let* ϕ *be a contractive linear map of* Y, *a linear subspace of* $\mathcal{B}$, *with* $1 \in Y$, *into a commutative* C*-*algebra then* ϕ *is completely contractive*.

Now let A be a function algebra with 1 and consider it as a closed subalgebra of $C(\partial A)$, where ∂A is the Šilov boundary of A, a compact Hausdorff space. A (contractive) representation of A is a homomorphism $\phi: A \to \mathcal{B}(H)$ such that $||\phi(f)|| \leq ||f||$, $(f \in A)$ and $\phi(1) = I$. By applying Stinespring's theorem (2.4) and Arveson's extension theorem (2.5, 2.6) to the C*-algebra $C(\partial A)$ and subspace A we obtain Arveson's dilation theorem [2, p. 278].

2.9 *Theorem*: *A representation* ϕ *of* A *in* $\mathcal{B}(H)$ *is completely contractive if and only if there exists a* *-*representation* ψ *of* $C(\partial A)$ *in* $\mathcal{B}(K)$, *some Hilbert space* K, *and a linear map* $V: H \to K$ *such that* $\phi(f) = V^*\psi(f)V$, $(f \in A)$ (*for* $f = 1$ *this shows* $I = V^*V$, *so that* V *is an isometry of* H *into* K).

The representation ψ and the map V are referred to as a *dilation* of ϕ. In brief, the theorem says exactly the completely contractive representations are dilatable. We are now ready to apply the concept of completely contractive representations to semigroups. For an index set θ, recall that A_θ is the uniform closure of P_θ, the polynomial functions in the complex variables $\{z_j : j \in \theta\}$ on the polydisc $\bar{U}^\theta$. Also A_θ is a function algebra with Silov boundary $\mathbb{T}^\theta$. For an $m \times m$ matrix $(f_{rs})_{r,s=1}^m$, $f_{rs} \in C(\mathbb{T}^\theta)$ let $||(f_{rs})||_\infty^{(m)}$ denote the C*-norm on $C(\mathbb{T}^\theta)^{(m)}$. Suppose ϕ is a representation of a semigroup S in $B(H)$, $x_j \in S$ for $j \in \theta$, and $p \in P_\theta^{(m)}$ (the subalgebra of $C(\mathbb{T}^\theta)^{(m)}$ with polynomial function entries), let $p(\phi x)$ denote the element of $B(H)^{(m)}$ which has the (rs)-entry given by $\Sigma_{\alpha \in Z_+^\theta} a_{rs,\alpha}\phi(x^\alpha)$ where $p_{rs} = \Sigma_\alpha a_{rs,\alpha} z^\alpha$ $(1 \le r,s \le m)$. Arveson's dilation theorem makes it possible to give an intrinsic characterization of type C representations.

2.10 *Theorem*: *Let* ϕ *be a type* A *representation of a semi-topological semigroup* S *in* $B(H)$. *Then* ϕ *is type* C *if and only if*

(2.10*) $$||p(\phi x)|| \le ||p||_\infty^{(m)}$$

for each $n = 1,2,3,\cdots$, $x_1,\cdots,x_n \in S$, $p \in P_n^{(m)}$, $m = 1,2,\cdots$. (*This property will be called being "completely* Q-*bounded"*.)

Proof. First suppose (2.10*) holds. Let $\{x_j\}_{j \in \theta}$ be a set of generators for S. Let ϕ_1 be the representation of Z_+^θ in $B_1(H)$ given by $\phi_1\alpha = \phi(x^\alpha)$ for $\alpha \in Z_+^\theta$. By hypothesis ϕ_1 induces a completely contractive representation $\phi_2 : A_\theta \to B(H)$, where

$\phi_2 p = p(\phi x)$ $(p \in P_\theta)$. By Arveson's Theorem 2.8 there exists a Hilbert space K containing H as a closed subspace and a $*$-representation ψ_1 of $C(\mathbb{T}^\theta)$ on $B(K)$ such that $\phi_2(f) = \mathrm{pr}\ \psi_1(f)|H$ for $f \in A_\theta$ (pr is the projection of K on H). There is an obvious representation of Z_+^θ in $C(\mathbb{T}^\theta)$, given by $\alpha \mapsto z^\alpha$, so define $\psi: Z_+^\theta \to B(K)$ by $\psi\alpha = \psi_1(z^\alpha)$ $(\alpha \in Z_+^\theta)$. We observe that $\psi\alpha$ is normal, indeed it is unitary since $|z^\alpha| = 1$ on $\mathbb{T}^\theta$. Finally, let $\rho: Z_+^\theta \to S$ be defined by $\rho\alpha = x^\alpha$, then $\mathrm{pr}(\psi\alpha)|H = \phi_2(z^\alpha) = \phi(x^\alpha)$ $= \phi(\rho\alpha)$ for $\alpha \in Z_+^\theta$. Thus ϕ is of type C, with Z_+^θ taking the part of S_0 in Definition 2.1.

Conversely let ϕ be of type C. We may as well assume that $S_0 = Z_+^\theta$ for some index set θ because the topology on S_0 is immaterial to the definition and any discrete semigroup is a quotient of some Z_+^θ. By hypothesis, there exists a homomorphism ρ of Z_+^θ onto S and a representation ψ of Z_+^θ in $B(K)$, where K is a Hilbert space $\supset H$, such that $\psi\alpha$ is normal and $\phi(\rho\alpha) = \mathrm{pr}\ \psi\alpha|H$ for $\alpha \in Z_+^\theta$.

Let A be the closed subalgebra of $B(K)$ generated by $\{\psi\alpha, \psi\alpha^* : \alpha \in Z_+^\theta\}$. The proof of Theorem 1.8 showed that A is a commutative C*-algebra. Now let $x_1, x_2, \cdots, x_n \in S$ for some $n = 1,2,\cdots$. Pick $\alpha_{(1)}, \cdots, \alpha_{(n)} \in Z_+^\theta$ such that $\rho\alpha_{(j)} = x_j$ $(1 \leq j \leq n)$. Thus we have a homomorphism $\pi: Z_+^n \to A \subset B(K)$ given by $\pi\beta = \Pi_{j=1}^n \psi(\alpha_{(j)})^{\beta_j}$, where $\beta = (\beta_1, \cdots, \beta_n) \in Z_+^n$. Since π is a normal representation of Z_+^n, Theorem 1.8 shows that π is Q-bounded, and thus $||\Sigma_{\beta \in Z_+^n} a_\beta \pi(\beta)|| \leq ||\Sigma_\beta a_\beta z^\beta||_\infty$ for each $\Sigma_\beta a_\beta z^\beta \in P_n$.

Now π induces a (contractive) homomorphism $\pi_1: A_n \to A \subset B(K)$ such that $\pi(\beta) = \pi_1(z^\beta)$, $\beta \in Z_+^n$. By Proposition 2.8, π_1 is completely contractive. Note that by construction

$\mathrm{pr}\ \pi_1(z^\beta)|H = \phi(x^\beta)$ for $\beta \in Z_+^n$. The projection to a subspace of a completely contractive linear map is still completely contractive, so $p \mapsto \mathrm{pr}\ \pi_1(p)|H = p(\phi x)$ is completely contractive ($p \in P_n$). Thus (2.10*) holds. □

2.11 <u>Corollary</u>: <u>If ϕ is type C then the semigroup S_0 in Definition 2.1 can be taken to be Z_+^θ for some index set θ and the representation ψ of Z_+^θ can be taken to consist of unitary operators.</u>

Proof. This statement is contained in the first part of the proof of Theorem 2.10. □

Type B does not imply type C. Parrott [1] has constructed three commuting contractions T_1, T_2, T_3 on a Hilbert space such that $||\Sigma_{\alpha \in Z_+^3} a_\alpha T^\alpha|| \leq ||\Sigma_\alpha a_\alpha z^\alpha||_\infty$ for each $\Sigma_\alpha a_\alpha z^\alpha \in P_3$ and for which there do <u>not</u> exist a Hilbert space $K \supset H$ and unitary operators $U_j \in B(K)$ so that $\mathrm{pr}\ U^\alpha|H = T^\alpha$ ($\alpha \in Z_+^3$). Thus $\alpha \mapsto T^\alpha$ ($\alpha \in Z_+^3$) is a type B representation of Z_+^3 which is not type C (invoking Corollary 2.11).

Some sufficient conditions are known for a type A representation of Z_+^θ to be dilatable to a unitary representation. These can be applied to semigroups.

2.12 <u>Theorem</u>: <u>Let ϕ be a type A representation of S and let $\{x_j\}_{j \in \theta}$ be a set of generators for S. Any of the following conditions imply that ϕ is of type C</u>:

(1) <u>Each ϕx_j, $j \in \theta$, is an isometry</u>;

(2) <u>For $j,k \in \theta$ with $j \neq k$</u>, $(\phi x_j)(\phi x_k)^* = (\phi x_k)^*(\phi x_j)$;

(3) $\Sigma_{j \in \theta}||\phi x_j||^2 \leq 1$.

The theorem is an immediate consequence of a theorem of Foiaş and Sz.-Nagy [1, p. 39]. A consequence of (3) is von Neumann's theorem that any type A representation of Z_+ is type B, indeed it is even type C. Arveson [1, p. 206] also proved this in the context of his theory, by showing that a contractive representation of a Dirichlet algebra is completely contractive. Note that A_1 is a Dirichlet algebra.

§3. Dilation theory for separative semigroups

We will have a brief look at the problem of extending (dilating) Hilbert space representations of closed subsemigroups of semigroups of type $\mathcal{U}$ to the whole semigroup.

3.1 Definition: Let S_0 be a semigroup of type $\mathcal{U}$ (dense inverse semigroup) and let S be a closed subsemigroup of S_0, with $1 \in S$. A type A representation ϕ of S in $\mathcal{B}(\mathcal{H})$ is said to be type D if there exists a type A representation ψ of S_0 on a Hilbert space $\mathcal{K} \supset \mathcal{H}$ such that $\phi x = \text{pr}(\psi x)|\mathcal{H}$ for $x \in S$.

3.2 Proposition: Each type D representation is also type C.

Proof. We use the notation of 3.1. Let U be a dense inverse semigroup in S_0. For each $x \in U$ we have $\psi x' = (\psi x)^*$ (see 3.1.5). So Theorem 1.8 applies to show that ψx is normal for each $x \in S_0$. Restrict ψ to S to obtain a normal representation of S in $\mathcal{B}(\mathcal{K})$ which projects onto ϕ, and thus ϕ is type C. □

3.3 Corollary: Let S be a semigroup of type $\mathcal{U}$, with identity 1, then any type A representation of S is also of types B, C, and D.

3.4 Example: Let S be a separative discrete semigroup with identity 1. The standard regular representation of S defined in 4.7.10 is a type D representation. It is the projection of a (normal) representation of an inverse semigroup $U \supset S$, namely, the projection of the regular representation on $\ell^2(U)$ to the subspace $\ell^2(S)$.

We point out that there exist compact semitopological semigroups of type $\mathcal{U}$ which do not have faithful Hilbert space representations. For example, let S be the weakly almost periodic compactification of a noncompact LCA group G, then $RQ(S) = R(S) \cong M(\hat{G})^{\wedge}$ and $R(S)$ is not uniformly dense in $C(S) = WAP(G)$ (see 5.2.10). By Corollary 3.3 and Theorem 1.8, any type A representation of S can separate only those points in S which are separated by $R(S)$, thus can not be faithful.

Use of the methods of Arveson yields a dilation theorem for normal representations of separative semigroups.

3.5 Theorem: Let S be a discrete separative semigroup and let U be a discrete inverse semigroup containing S. Let ϕ be a type A representation of S by normal operators, in $B(H)$, then ϕ is type D.

Proof. Let A be the closed subalgebra of $B(H)$ generated by $\{\phi x, \phi x^* : x \in S\}$, then A is a commutative C*-algebra. Let $\xi, \eta \in H$, then $x \mapsto \langle(\phi x)\xi,\eta\rangle$ is in $R(S)$ (these statements were proved in Theorem 1.8). From Theorem 4.5.1 we see that $|\Sigma_x a_x \langle(\phi x)\xi,\eta\rangle| \le |\xi||\eta| \, ||\Sigma_x a_x \delta_x||_{sp}$ for each finitely supported measure $\Sigma_x a_x \delta_x$. We thus have that $||\Sigma_x a_x \phi x|| \le ||\Sigma_x a_x \hat{x}||_{\hat{U}}$ (the spectral norm on $\ell^1(S)$ agrees with the sup-norm of the Gelfand transform over $\hat{U}$, see 4.5.1).

Let Y be the closed subalgebra of $C(\hat{U})$ generated by $\{\hat{x}:x \in S\}$. We see that ϕ induces a contractive representation of Y in A, a commutative C*-algebra, hence ϕ is completely contractive (2.8). Apply Theorem 2.9 to obtain a *-representation ψ of $C(\hat{U})$ in $B(K)$, where K is a Hilbert space $\supset H$ and such that pr $\psi(f)|H = \phi(f)$ $(f \in Y)$. Restrict ψ to $\{\hat{y}:y \in U\}$ to get a type A representation of U which dilates ϕ. □

No doubt more results remain to be discovered. A promising area would be dilation theory for a closed subsemigroup S of a locally compact abelian group G with S being the closure of its G-interior (e.g., $\mathbb{R}_+$ in $\mathbb{R}$). Type C would certainly be a necessary condition, as well as some knowledge about RQ(S) (the best thing would be $RQ(S) = RQ(G)|S = R(G)|S$, with the quotient norm). One such case, $\mathbb{R}_+$, has of course been long studied, under the names "semigroups of operators" and "one-parameter semigroups". We present those highlights of the theory which are relevant to our considerations.

Suppose ϕ is a type A representation of $\mathbb{R}_+$ on a Hilbert space H. Thus $\phi t \in B(H)$, $\|\phi t\| \leq 1$, $\phi(s+t) = (\phi s)(\phi t)$ and $\phi(0) = I$ $(s,t \geq 0)$. Further, $\langle(\phi s)\xi,\eta\rangle \to \langle(\phi s_0)\xi,\eta\rangle$ as $s \to s_0$ in $\mathbb{R}_+$, $(\xi,\eta \in H)$. In this case WO-continuity implies SO-continuity. Indeed let $t_2 > t_1 \geq 0$, put $s = t_2-t_1$, and let $\xi \in H$, then

$$|\phi(t_2)\xi-\phi(t_1)\xi|^2 = |\phi(t_1)((\phi s)\xi-\xi)|^2 \leq |(\phi s)\xi-\xi|^2$$
$$= |(\phi s)\xi|^2-2\mathrm{Re}\langle(\phi s)\xi,\xi\rangle+|\xi|^2 \leq 2|\xi|^2-2\mathrm{Re}\langle(\phi s)\xi,\xi\rangle = 2\mathrm{Re}\langle\xi-(\phi s)\xi,\xi\rangle.$$

The last term tends to 0 as $s \to 0+$, and so $t \to \phi t$ is SO (strong operator) continuous. For each $h > 0$ define an operator

$A_h \in B(H)$ by $A_h\xi = \frac{1}{h}(\phi(h)\xi - \xi)$, $(\xi \in H)$. Then define

$A\xi = \lim_{h\to 0+} A_h\xi$ for those ξ for which the limit exists, in the norm of H. Then A is a closed linear operator with dense domain in H, and it is called the infinitesimal generator. In addition ϕ is norm continuous, that is, $||\phi s - \phi s_0|| \to 0$ as $s \to s_0$ in $\mathbb{R}_+$, if and only if A is bounded and defined on all of H. Formally, $\phi t = \exp tA$. This formula is realized as

$$(\phi t)\xi = \lim_{h\to 0} (\exp tA_h)\xi, \quad \text{for each } \xi \in H .$$

The proofs may be found in Rudin [2, pp. 355-360] for example. Since ϕ is a contractive representation, it is further known that $||(I-n^{-1}A)^{-1}|| \leq 1$, for $n = 1,2,3,\cdots$ where $(I-n^{-1}A)^{-1}$ is the resolvent of A, an element of $B(H)$ (see Yosida [1, p. 249]).

If each ϕt is normal, then A is a normal operator with spectrum contained in $\{z \in \mathbb{C}: \text{Re } z \leq 0\}$. If each ϕt is unitary, then $A = iH$, where H is a Hermitian operator (this is a classical theorem of M.H. Stone) (see Rudin [2, p. 360]).

3.6 Theorem: Let ϕ be a type A representation of $\mathbb{R}_+$ (additive) on a Hilbert space H, then there exists a unitary representation ψ of $\mathbb{R}$ on a Hilbert space $K \supset H$ such that $\phi s = \text{pr } \psi s$ $(s \geq 0)$. Also K is spanned by $(\psi t)\xi$ $(t \in \mathbb{R}, \xi \in H)$, a minimality condition.

3.7 Corollary: Any type A representation of $\mathbb{R}_+$ is also type D.

8.3.7

The theorem may be found in Foiaş and Sz.-Nagy [1, p. 31]. The proof depends on the Z_+-theory so is somewhat ad hoc. The examples of Varopoulos and Parrott show that one would not expect the analogous theorem for $\mathbb{R}_+^n$, $n \geq 3$, but perhaps a completely contractive (type C) condition may imply the existence of a dilation.

Appendix A. Fourier Analysis

We give here some of the basic facts of Fourier analysis which can be found in Rudin [1]. On every locally compact abelian (LCA) group G there exists a positive regular Borel measure which is translation-invariant. It is called the Haar measure of G and is denoted m_G. The space M(G) is the collection of all finite regular Borel measures on G. It is a commutative Banach algebra with unit under the operation of convolution *: for $\mu, \nu \in M(G)$, $\mu * \nu \in M(G)$ is defined by

$$\int_G f d\mu * \nu = \int_G \int_G f(x+y)\, d\mu(x)\, d\nu(y), \quad f \in C_o(G).$$

Recall the dual space $C_o(G)^*$ of $C_o(G)$ is M(G) (Riesz representation theorem). The subspace $L^1(G)$ consists of all $\nu \in M(G)$ which are absolutely continuous with respect to m_G ($\nu << m_G$). It can be identified with the space of equivalence classes of Borel functions f on G with $\int_G |f| dm_G < \infty$.

The dual (or character) group $\hat{G}$ of G is the set of all continuous homomorphisms of G into the unit circle $\mathbb{T}$ in the complex plane $\mathbb{C}$. The topology of $\hat{G}$ is the compact-open topology; and G is also an LCA group. The space G can be identified with the maximal ideal space of $L^1(G)$ with the Gelfand topology. The Pontryagin duality theorem asserts that $G^{\wedge\wedge} = G$.

For $\mu \in M(G)$, the Fourier-Stieltjes transform $\hat{\mu}$ is a continuous, bounded function on $\hat{G}$ and is defined by

$$\hat{\mu}(\gamma) = \int_G \gamma(x)\, d\mu(x), \quad \gamma \in \hat{G}.$$

The map $\mu \mapsto \hat{\mu}: M(G) \to C^B(\hat{G})$ is a faithful, norm nonincreasing, algebra homomorphism. The Bochner theorem asserts that a function f on $\hat{G}$ is a continuous, positive-definite function if and only if $f = \hat{\mu}$, some $\mu \in M(G)$ with $\mu \geq 0$.

The space $M(G)\hat{}$ has been characterized by Eberlein as the space of functions $f \in C^B(\hat{G})$ with the property

$$|\Sigma_{i=1}^n c_i f(\gamma_i)| \leq K \sup\{|\Sigma_{i=1}^n c_i \gamma_i(x)| : x \in G\}$$

for all $c_1,\cdots,c_n \in \mathbb{C}$, $\gamma_1,\cdots,\gamma_n \in \hat{G}$ (K a constant depending only on f).

Appendix B. Spectral Theorem

This is a version of the spectral theorem adequate for our purposes, see Rudin [2] and Naimark [1].

For the σ-algebra M of all Borel sets on a locally compact Hausdorff space Ω, a resolution of the identity is a mapping E of M into the bounded operators $B(H)$ on some Hilbert space H with inner product $\langle\cdot,\cdot\rangle$ such that:

(a) $E(\emptyset) = 0$, $E(\Omega) = I$

(b) $E(\omega)$ is a self-adjoint projection $(\omega \in M)$

(c) $E(\omega_1 \cap \omega_2) = E(\omega_1)E(\omega_2) \quad (\omega_1, \omega_2 \in M)$

(d) if $\omega_1 \cap \omega_2 = \emptyset$, then

$$E(\omega_1 \cup \omega_2) = E(\omega_1) + E(\omega_2) \quad (\omega_1, \omega_2 \in M)$$

(e) for each $x, y \in H$, the set function $E_{x,y}$ defined by $E_{x,y}(\omega) = \langle E(\omega)x, y\rangle$, $(\omega \in M)$ is a finite regular Borel measure on M.

For a resolution of the identity E on M, the Banach algebra $L^\infty(E)$ is the space of equivalence classes of all bounded complex M-measurable functions on Ω with the essential sup-norm (the equivalence relation is defined by $f \sim g$ if and only if $E(\{\omega \in \Omega : f(\omega) \neq g(\omega)\}) = 0$.

Given a resolution of the identity on M, there exists a isometric $*$-isomorphism ψ of $L^\infty(E)$ onto a closed normal subalgebra A of $B(H)$ given by

$$(*) \qquad \langle\psi(f)x, y\rangle = \int_\Omega f\,dE_{x,y} \quad (f \in L^\infty(E), x, y \in H).$$

We abbreviate $(*)$ to $\psi(f) = \int_\Omega f\,dE$.

Let A be a closed normal subalgebra of $\mathcal{B}(H)$ which contains the identity operator I. Let Δ_A denote the maximal ideal space (Δ is compact) of A. Then there exists a unique resolution of the identity E on the Borel subsets of Δ_A with

$$(**) \qquad \langle Sx,y\rangle = \int_{\Delta_A} \hat{S}dE_{x,y} ,$$

where $S \mapsto \hat{S}$ denotes the Gel'fand transform of A onto $C(\Delta_A)$. We abbreviate (**) by writing $S = \int_{\Delta_A} \hat{S}dE$.

Let A be a commutative C*-algebra with a unit, and let $f \mapsto T_f : A \to \mathcal{B}(H)$ be a continuous $*$-representation of A. The representation T is said to by <u>cyclic</u> if there exists a vector $x_o \in H$ with $||x_o|| = 1$ and the set $\{T_f(x_o) : f \in A\}$ is dense in H. Let p be the positive functional on A defined by x_o; that is $p(f) = \langle T_f(x_o), x_o\rangle$. By the Riesz representatinn theorem, there exists a measure $\mu \in M_p(\Delta_A)$ with

$$p(f) = \int_{\Delta_A} \hat{f}d\mu, \quad (f \in A).$$

For $f \in A$, let $j(T_f(x_o))$ be the function $\hat{f} \in C(\Delta_A) \subset L^2(\mu)$. The map j extends to a linear isometry of H onto $L^2(\mu)$; and for $\phi \in L^2(\mu)$, $jT_fj^{-1}(\phi) = \hat{f}\,\phi(f \in A)$. Thus T is equivalent to the representation $f \mapsto \hat{f} : A \to L^\infty(\mu) \subset \mathcal{B}(L^2(\mu))$.

Appendix C. The structure semigroup of the representation algebra

In this section the semigroup S is a commutative semitopological semigroup with identity 1. We will characterize the dual $R(S)^*$ $(=A(S))$ of $R(S)$.

1 Definition: For $f \in R(S)$ and $F \in R(S)^*$, define the function $E_F f$ on S by $E_F f(y) = F(f_y)$ $(y \in S)$ (recall 2.1.11).

The following is a crucial proposition.

2 Proposition: The operator $E_F: f \mapsto E_F f$ takes $R(S)$ into $R(S)$ with $||E_F|| \leq ||F||$ $(F \in R(S)^*)$.

Proof. Fix $f \in R(S)$. Recall from Chapter 2 (2.1.10), the spaces $W = \Sigma \oplus_{T \in S} L^\infty(\mu,\Omega)$ and $L = \Sigma \oplus_{T \in S} L^1(\mu,\Omega)$ $((T,\mu,\Omega) \in S)$ with $L^* = W$. For $w \in W$, $g \in L$, the canonical pairing between W and L is $\langle w,g\rangle = \Sigma_{T \in S} \int_\Omega w_T g\, d\mu$. For $x \in S$, $\rho x \in W$ is defined by $(\rho x)_T = Tx$ $(T \in S)$. For $x \in S$, $\rho^*: L \to R(S)$ is defined by $\rho^* g(x) = \langle \rho x, g\rangle$ $(g \in L)$. By Theorem 1.1.12, $L^*/\ker \rho^* = R(S)$. Thus for a fixed $f \in R(S)$, let $g \in L$ with $\rho^* g = f$ and $||f||_R = ||g||_L$, so that $f(z) = \rho^* g(z) = \langle \rho z, g\rangle$ $(z \in S)$. For $h \in L$ and $w \in W$, define $h \times w \in L$ by $h \times w = (w_T h_T)_{T \in S}$. Thus for $h \in L$ and $w_1, w_2 \in W$, $\langle w_1, h \times w_2\rangle = \langle w_2, h \times w_1\rangle = \langle w_1 w_2, h\rangle$; in particular, $\langle w(\rho y), g\rangle = \Sigma_{T \in S} \int_\Omega w_T(Ty) g_T d\mu = \langle w, g \times (\rho y)\rangle$ $(y \in S)$. Hence $f_x(y) = f(xy) = \langle \rho(xy), g\rangle = \langle (\rho x)(\rho y), g\rangle = \langle \rho y, g \times (\rho x)\rangle$ $= \rho^*(g \times (\rho x))(y)$, $(x,y \in S)$; that is, $f_x = \rho^*(g \times \rho(x))$.

Each $F \in R(S)^*$ defines a bounded linear functional $F^\# \in L^* = W$ by $\langle F^\#, g\rangle = F(\rho^* g)$, $(g \in L)$. So for $x \in S$, $(E_F f)(x) = F(f_x)$

$= F(\rho^*(g^\times(\rho x))) = \langle F^{\#}, g^\times(\rho x)\rangle = \langle \rho x, g^\times F^{\#}\rangle = \rho^*(g^\times F^{\#})(x)$; and so $E_F f \in R(S)$. Also $|||E_F f||_{R(S)} \leq ||\rho^*||\ ||g^\times F^{\#}||_L$ $\leq ||g||_L ||F|| = ||f||_R ||F||$; so $||E_F|| \leq ||F||$. □

3 <u>Definition</u>: Let $B_S(R(S))$ denote the bounded operators on $R(S)$ which commute with translation; that is for $\Phi \in B_S(R(S))$, $\Phi f_x = (\Phi f)_x$ $(f \in R(S),\ x \in S)$.

4 <u>Theorem</u>: <u>Let</u> S <u>be</u> <u>a</u> <u>commutative</u> <u>semitopological</u> <u>semigroup</u> <u>with</u> <u>identity</u> 1. <u>The</u> <u>map</u> $F \mapsto E_F$ <u>is</u> <u>a</u> <u>one-to-one</u> <u>linear</u> <u>isometry</u> <u>of</u> $R(S)^*$ <u>onto</u> $B_S(R(S))$.

Proof. Let $F \in R(S)^*$. Then $||E_F|| \leq ||F||$, and $|F(f)| = |E_F(f_1)| \leq ||E_F||\ ||f_1||_R = ||E_F||\ ||f||_R$. Thus $||E_F|| = ||F||$; that is, $F \mapsto E_F$ is a linear isometry of $R(S)^*$ into $B(R(S))$. Also $E_F \in B_S(R(S))$: for $x,y \in S$, $(E_F f)_x(y) = E_F f(xy)$ $= F(f_{xy}) = F((f_x)_y) = E_F(f_x)(y)$, $(f \in R(S))$.

For $E \in B_S(R(S))$, define $F \in R(S)^*$ by $F(f) = E(f)(1)$, $(f \in R(S))$. Now $E_F = E$: for $f \in R(S)$, $(E_F f)(x) = F(f_x) = E(f_x)(1)$ $= E(f)(x)$, $(x \in S)$. □

5 <u>Corollary</u>: <u>Let</u> S <u>be</u> <u>a</u> <u>commutative</u> <u>semitopological</u> <u>semigroup</u> <u>with</u> <u>identity</u> 1. <u>The</u> <u>maximal</u> <u>ideal</u> <u>space</u> Δ <u>of</u> $R(S)$ <u>is</u> <u>identified</u> <u>with</u> <u>the</u> <u>nonzero</u> <u>endomorphisms</u> <u>in</u> $B_S(R(S))$.

Proof. The composition of two nonzero endomorphisms in nonzero since $1 \in S$. □

6 Corollary: *Let S be a commutative semitopological semigroup with identity 1. The space $R(S)^*$ with the weak-* topology is homeomorphic to the space $B_S(R(S))$ with the weak operator topology. Thus Δ is a compact semitopological semigroup with an identity.*

Proof. Recall from 2.1.13 that $R(S)^* = A(S) \subset W$ is isomorphic to the weak-* closed span of $\{\rho x : x \in S\}$ in W. For $\{F_\alpha\} \cup \{F\} \subset R(S)^*$, $E_{F_\alpha} \overset{\alpha}{\to} E_F$ WO if and only if $E_{F_\alpha} f \overset{\alpha}{\to} E_F f$ weakly in $R(S)$ $(f \in R(S))$ if and only if $\langle w, E_{F_\alpha} f\rangle \overset{\alpha}{\to} \langle w, E_F f\rangle$ $(w \in A(S),\ f \in R(S))$ if and only if $\langle w, g \times F_\alpha^{\#}\rangle \overset{\alpha}{\to} \langle w, g \times F^{\#}\rangle$ $(g \in L,\ w \in A(S))$ if and only if $\langle F_\alpha^{\#}, g \times w\rangle \overset{\alpha}{\to} \langle F^{\#}, g \times w\rangle$ $(g \in L,\ w \in A(S))$ if and only if $\langle F_\alpha^{\#}, g\rangle \overset{\alpha}{\to} \langle F^{\#}, g\rangle$ $(g \in L)$ (since $L \times A(S) = L$) if and only if $F_\alpha \overset{\alpha}{\to} F$ weak-* in $R(S)$. □

7 Remark: For S a locally compact abelian group, these results are due to Taylor [2].

References

Andó, T.

1. On a pair of commutative contractions. Acta. Sci. Math. 24 (1963), 88-90.

Arens, R.

1. The adjoint of a bilinear operation. Proc. Amer. Math. Soc. 2 (1951), 839-848.

Arens, R. and Singer, I.

1. Generalized analytic functions. Trans. Amer. Math. Soc. 81 (1956), 379-393.

Arveson, W.

1. Subalgebras of C*-algebras. Acta. Math. 123 (1969), 141-224.

2. ______________________________, II. Acta. Math. 128 (1972), 271-308.

Austin, C.

1. Duality theorems for some commutative semigroups. Trans. Amer. Math. Soc. 109 (1963), 245-256.

Berglund, J. and Hofmann, H.

1. Compact Semitopological Semigroups and Weakly Almost Periodic Functions. Springer-Verlag, Berlin, 1967.

Bonsall, F. and Duncan, J.

1. Complete Normed Algebras. Springer-Verlag, Berlin, 1973.

Brown, D. and Friedberg, M.

1. A new notion of semicharacters. Trans. Amer. Math. Soc. 141 (1969), 387-401.

Brown, G. and Moran, W.

1. Idempotents of compact monothetic semigroups. Proc. London Math. Soc. 22 (1971), 211-221.

Burckel, R.

1. Weakly Almost Periodic Functions on Semigroups. Gordon and Breach, New York, 1970.

Civin, P. and Yood, B.

1. The second conjugate space of a Banach algebra as an algebra. Pacific J. Math. 3 (1961), 847-870.

Comfort, W.

1. The Silov boundary induced by a certain Banach algebra. Trans. Amer. Math. Soc. 98 (1961), 501-517.

Davie, A.

1. Quotient algebras of uniform algebras. J. London Math. Soc. (2), 7 (1973), 31-40.

deLeeuw, K. and Glicksberg, I.

1. Applications of almost periodic compactifications. Acta. Math. 105 (1961), 99-140.

2. The decomposition of certain group representations. J. d'Analyse Math. 40 (1965), 135-192.

Dixmier, J.

1. Sur certains espaces consideres par M.H. Stone. Summa Brasil. Math. 2 (1951), 151-182.

Doss, R.

1. Approximations and representations for Fourier transforms. Trans. Amer. Math. Soc. 153 (1971), 211-221.

Dunford, N. and Schwartz, T.

1. Linear Operators I. Interscience, New York, 1958.

Dunkl, C.

1. A semigroup-type definition of Fourier-Stieltjes transforms. Notices Amer. Math. Soc. 21 (1974), A-166.

2. Bounded derivations on commutative semigroups.

Dunkl, C. and Ramirez, D.

1. Topics in Harmonic Analysis. Appleton-Century-Crofts, New York, 1971.

2. L^∞-representations of commutative semitopological semigroups. Semigroup Forum 7 (1974), 180-199.

3. Sections induced from weakly sequentially complete spaces. Studia Math. 49 (1973), 97-99.

Eberlein, W.

1. Abstract ergodic theorems and weak almost periodic functions. Trans. Amer. Math. Soc. 67 (1949), 217-240.

Edwards, R.

1. On functions which are Fourier transforms. Proc. Amer. Math. Soc. 5 (1954), 71-78.

Ellis, R.

1. Locally compact transformation groups. Duke Math. J. 24 (1967), 119-126.

Fine, N. and Maserick, P.

1. On the simplex of completely monotonic functions on a commutative semigroup. Canad. J. Math. 22 (1970),317-326.

Foias, C. and Sz.-Nagy, B.

1. _Harmonic Analysis of Operators on Hilbert Space_. North-Holland, Amsterdam, 1970.

Freyd, P.

1. Redei's finiteness theorem for commutative semigroups. Proc. Amer. Math. Soc. 19 (1968), 1003.

Gamelin, T.

1. _Uniform Algebras_. Prentice-Hall, New Jersey, 1969.

Gelfand, I. and Naimark, M.

1. On the imbedding of normed rings into the ring of operators in Hilbert space. Mat. Sbornik N.S. 12 (1943), 197-213.

Glicksberg, I.

(see deLeeuw)

Hewitt, E. and Ross, K.

1. _Abstract Harmonic Analysis I_. Springer-Verlag, Berlin, 1963.

Hewitt, E. and Zuckerman, H.

1. Finite dimensional convolution algebras. Acta. Math. 93 (1955), 67-119.

2. The ℓ^1-algebra of a commutative semigroup. Trans. Amer. Math. Soc. 83 (1956), 70-97.

Hofmann, K. (see Berglund)

1. The Duality of Compact Semigroups and C*-Bigebras, Lecture Notes in Mathematics v. 129. Springer-Verlag, Berlin, 1970.

Hofmann, K. and Mostert, P.

1. Elements of Compact Semigroups. Charles E. Merrill Books, Columbus, 1966.

Holbrook, J.

1. Spectral dilations and polynomially bounded operators. Indiana Math. J. 20 (1971), 1027-1034.

Kaplansky, I.

1. A theorem on rings of operators. Pacific J. Math. 1 (1951), 227-232.

Köthe, G.

1. Topological Vector Spaces I. Springer-Verlag, New York, 1969.

Lawson, J.

1. Joint continuity in semitopological semigroups. Illinois J. Math. 18 (1974), 275-285.

Lindahl, R. and Maserick, P.

1. Positive-definite functions on involution semigroups. Duke Math. J. 38 (1971), 771-782.

Maserick, P. (see Lindahl)

1. Moment and BV-functions on commutative semigroups. Trans. Amer. Math. Soc. 181 (1973), 61-75.

Moran, W.

(see G. Brown)

Mostert, P.

(see Hofmann)

Naimark, M. (see Gelfand)

1. Normed Rings. P. Noordhoff, N.V., Groningen, The Netherlands, 1964.

Newman, S.

1. Measure algebras and functions of bounded variation on idempotent semigroups. Bull. Amer. Math. Soc. 75 (1969), 1396-1400.

Nussbaum, A.

1. The Hausdorff-Bernstein-Widder theorem for semigroups in locally compact abelian groups. Duke Math. J. 22 (1955), 573-582.

Parrott, S.

1. Unitary dilations for commuting contractions. Pacific J. Math. 34 (1970), 481-490.

Ramirez, D.

(see Dunkl)

Redéi, L.

1. The Theory of Finitely Generated Commutative Semigroups. Pergamon Press, New York, 1965.

Rosenthal, H.

1. A characterization of Fourier-Stieltjes transforms. Pacific J. Math. 23 (1967), 403-418.

Ross, K. (see Hewitt)

1. A note on extending semicharacters on semigroups. Proc. Amer. Math. Soc. 10 (1959), 579-583.

Rudin, W.

1. Fourier Analysis on Groups. Interscience Publishers, New York, 1962.

2. Functional Analysis. McGraw Hill, New York, 1973.

3. Function Theory in Polydiscs. Benjamin, New York, 1969.

4. Weak almost periodic functions and Fourier-Stieltjes transforms. Proc. Amer. Math. Soc. 26 (1959), 215-220.

Sakai, S.

1. C*-Algebras and W*-Algebras. Springer-Verlag, Berlin, 1971.

2. On topological properties of W*-algebras. Proc. Japan Acad. 33 (1957), 439-444.

Saworotnow, P.

1. Semigroups with positive definite structure. Proc. Amer. Math. Soc. 40 (1973), 421-425.

Schwartz, J.

(see Dunford)

Schwarz, S.

1. The theory of characters of commutative Hausdorff bicompact semigroups. Czech. Math. J. 6 (1956), 330-361.

Seever, G.

1. Algebras of continuous functions on hyperstonian spaces. Arch. Math. 24 (1973), 648-660.

Shohat, J. and Tamarkin, J.

1. The problem of moments. Amer. Math. Soc. Math. Surveys 1, Providence, 1943.

Singer, I.

(see Arens)

Stinespring, W.

1. Positive functions on C*-algebras. Proc. Amer. Math. Soc. 6 (1955), 211-216.

Stout, E.

1. The Theory of Uniform Algebras. Bogden & Quigley, Inc. 1971.

Sz.-Nagy, B.

(see Foias)

Tamarkin, J.

(see Shohat)

Taylor, J.

1. Measure algebras. Amer. Math. Soc. regional conference series 16, Providence, 1972.

2. The structure of convolution measure algebras. Trans. Amer. Math. Soc. 119 (1965), 150-166.

Varopoulos, N.

1. On an inequality of von Neumann and an application of the metric theory of tensor products to operator theory. J. Functional Analysis 16 (1974), 83-100.

von Neumann, J.

1. Eine Spektraltheorie für allgemeine Operatoren eines unitären Raumes. Math. Nachr. 4 (1951), 258-281.

Warne, R. and Williams, L.

1. Characters on inverse semigroups. Czech. Math. J. 11 (1961), 150-154.

West, T.

1. Weakly compact monothetic semigroups of operators in Banach spaces. Proc. Royal Irish Acad. 67 (1968), 27-37.

Williams, L.

(see Warne)

Yood, B.

(see Civin)

Yosida, K.

1. Functional Analysis. 2nd ed.. Springer-Verlag, Berlin, 1968.

Young, N.

1. Semigroup algebras having regular multiplication. Studia Math. 47 (1973), 191-196.

Zuckerman, H.

(see Hewitt)

Zygmund, A.

1. Trigonometric Series I. Cambridge University Press, New York, 1959.

Symbol Index

Author Index

Subject Index

Vol. 277: Séminaire Banach. Edité par C. Houzel. VII, 229 pages. 1972. DM 20,–

Vol. 278: H. Jacquet, Automorphic Forms on GL(2). Part II. XIII, 142 pages. 1972. DM 16,–

Vol. 279: R. Bott, S. Gitler and I. M. James, Lectures on Algebraic and Differential Topology. V, 174 pages. 1972. DM 18,–

Vol. 280: Conference on the Theory of Ordinary and Partial Differential Equations. Edited by W. N. Everitt and B. D. Sleeman. XV, 367 pages. 1972. DM 26,–

Vol. 281: Coherence in Categories. Edited by S. Mac Lane. VII, 235 pages. 1972. DM 20,–

Vol. 282: W. Klingenberg und P. Flaschel, Riemannsche Hilbertmannigfaltigkeiten. Periodische Geodätische. VII, 211 Seiten. 1972. DM 20,–

Vol. 283: L. Illusie, Complexe Cotangent et Déformations II. VII, 304 pages. 1972. DM 24,–

Vol. 284: P. A. Meyer, Martingales and Stochastic Integrals I. VI, 89 pages. 1972. DM 16,–

Vol. 285: P. de la Harpe, Classical Banach-Lie Algebras and Banach-Lie Groups of Operators in Hilbert Space. III, 160 pages. 1972. DM 16,–

Vol. 286: S. Murakami, On Automorphisms of Siegel Domains. V, 95 pages. 1972. DM 16,–

Vol. 287: Hyperfunctions and Pseudo-Differential Equations. Edited by H. Komatsu. VII, 529 pages. 1973. DM 36,–

Vol. 288: Groupes de Monodromie en Géométrie Algébrique. (SGA 7 I). Dirigé par A. Grothendieck. IX, 523 pages. 1972. DM 50,–

Vol. 289: B. Fuglede, Finely Harmonic Functions. III, 188. 1972. DM 18,–

Vol. 290: D. B. Zagier, Equivariant Pontrjagin Classes and Applications to Orbit Spaces. IX, 130 pages. 1972. DM 16,–

Vol. 291: P. Orlik, Seifert Manifolds. VIII, 155 pages. 1972. DM 16,–

Vol. 292: W. D. Wallis, A. P. Street and J. S. Wallis, Combinatorics: Room Squares, Sum-Free Sets, Hadamard Matrices. V, 508 pages. 1972. DM 50,–

Vol. 293: R. A. DeVore, The Approximation of Continuous Functions by Positive Linear Operators. VIII, 289 pages. 1972. DM 24,–

Vol. 294: Stability of Stochastic Dynamical Systems. Edited by R. F. Curtain. IX, 332 pages. 1972. DM 26,–

Vol. 295: C. Dellacherie, Ensembles Analytiques, Capacités, Mesures de Hausdorff. XII, 123 pages. 1972. DM 16,–

Vol. 296: Probability and Information Theory II. Edited by M. Behara, K. Krickeberg and J. Wolfowitz. V, 223 pages. 1973. DM 20,–

Vol. 297: J. Garnett, Analytic Capacity and Measure. IV, 138 pages. 1972. DM 16,–

Vol. 298: Proceedings of the Second Conference on Compact Transformation Groups. Part 1. XIII, 453 pages. 1972. DM 32,–

Vol. 299: Proceedings of the Second Conference on Compact Transformation Groups. Part 2. XIV, 327 pages. 1972. DM 26,–

Vol. 300: P. Eymard, Moyennes Invariantes et Représentations Unitaires. II. 113 pages. 1972. DM 16,–

Vol. 301: F. Pittnauer, Vorlesungen über asymptotische Reihen. VI, 186 Seiten. 1972. DM 18,–

Vol. 302: M. Demazure, Lectures on p-Divisible Groups. V, 98 pages. 1972. DM 16,–

Vol. 303: Graph Theory and Applications. Edited by Y. Alavi, D. R. Lick and A. T. White. IX, 329 pages. 1972. DM 26,–

Vol. 304: A. K. Bousfield and D. M. Kan, Homotopy Limits, Completions and Localizations. V, 348 pages. 1972. DM 26,–

Vol. 305: Théorie des Topos et Cohomologie Etale des Schémas. Tome 3. (SGA 4). Dirigé par M. Artin, A. Grothendieck et J. L. Verdier. VI, 640 pages. 1973. DM 50,–

Vol. 306: H. Luckhardt, Extensional Gödel Functional Interpretation. VI, 161 pages. 1973. DM 18,–

Vol. 307: J. L. Bretagnolle, S. D. Chatterji et P.-A. Meyer, Ecole d'été de Probabilités: Processus Stochastiques. VI, 198 pages. 1973. DM 20,–

Vol. 308: D. Knutson, λ-Rings and the Representation Theory of the Symmetric Group. IV, 203 pages. 1973. DM 20,–

Vol. 309: D. H. Sattinger, Topics in Stability and Bifurcation Theory. VI, 190 pages. 1973. DM 18,–

Vol. 310: B. Iversen, Generic Local Structure of the Morphisms in Commutative Algebra. IV, 108 pages. 1973. DM 16,–

Vol. 311: Conference on Commutative Algebra. Edited by J. W. Brewer and E. A. Rutter. VII, 251 pages. 1973. DM 22,–

Vol. 312: Symposium on Ordinary Differential Equations. Edited by W. A. Harris, Jr. and Y. Sibuya. VIII, 204 pages. 1973. DM 22,–

Vol. 313: K. Jörgens and J. Weidmann, Spectral Properties of Hamiltonian Operators. III, 140 pages. 1973. DM 16,–

Vol. 314: M. Deuring, Lectures on the Theory of Algebraic Functions of One Variable. VI, 151 pages. 1973. DM 16,–

Vol. 315: K. Bichteler, Integration Theory (with Special Attention to Vector Measures). VI, 357 pages. 1973. DM 26,–

Vol. 316: Symposium on Non-Well-Posed Problems and Logarithmic Convexity. Edited by R. J. Knops. V, 176 pages. 1973. DM 18,–

Vol. 317: Séminaire Bourbaki – vol. 1971/72. Exposés 400–417. IV, 361 pages. 1973. DM 26,–

Vol. 318: Recent Advances in Topological Dynamics. Edited by A. Beck, VIII, 285 pages. 1973. DM 24,–

Vol. 319: Conference on Group Theory. Edited by R. W. Gatterdam and K. W. Weston. V, 188 pages. 1973. DM 18,–

Vol. 320: Modular Functions of One Variable I. Edited by W. Kuyk. V, 195 pages. 1973. DM 18,–

Vol. 321: Séminaire de Probabilités VII. Edité par P. A. Meyer. VI, 322 pages. 1973. DM 26,–

Vol. 322: Nonlinear Problems in the Physical Sciences and Biology. Edited by I. Stakgold, D. D. Joseph and D. H. Sattinger. VIII, 357 pages. 1973. DM 26,–

Vol. 323: J. L. Lions, Perturbations Singulières dans les Problèmes aux Limites et en Contrôle Optimal. XII, 645 pages. 1973. DM 42,–

Vol. 324: K. Kreith, Oscillation Theory. VI, 109 pages. 1973. DM 16,–

Vol. 325: Ch.-Ch. Chou, La Transformation de Fourier Complexe et L'Equation de Convolution. IX, 137 pages. 1973. DM 16,–

Vol. 326: A. Robert, Elliptic Curves. VIII, 264 pages. 1973. DM 22,–

Vol. 327: E. Matlis, 1-Dimensional Cohen-Macaulay Rings. XII, 157 pages. 1973. DM 18,–

Vol. 328: J. R. Büchi and D. Siefkes, The Monadic Second Order Theory of All Countable Ordinals. VI, 217 pages. 1973. DM 20,–

Vol. 329: W. Trebels, Multipliers for (C, α)-Bounded Fourier Expansions in Banach Spaces and Approximation Theory. VII, 103 pages. 1973. DM 16,–

Vol. 330: Proceedings of the Second Japan-USSR Symposium on Probability Theory. Edited by G. Maruyama and Yu. V. Prokhorov. VI, 550 pages. 1973. DM 36,–

Vol. 331: Summer School on Topological Vector Spaces. Edited by L. Waelbroeck. VI, 226 pages. 1973. DM 20,–

Vol. 332: Séminaire Pierre Lelong (Analyse) Année 1971-1972. V, 131 pages. 1973. DM 16,–

Vol. 333: Numerische, insbesondere approximationstheoretische Behandlung von Funktionalgleichungen. Herausgegeben von R. Ansorge und W. Törnig. VI, 296 Seiten. 1973. DM 24,–

Vol. 334: F. Schweiger, The Metrical Theory of Jacobi-Perron Algorithm. V, 111 pages. 1973. DM 16,–

Vol. 335: H. Huck, R. Roitzsch, U. Simon, W. Vortisch, R. Walden, B. Wegner und W. Wendland, Beweismethoden der Differentialgeometrie im Großen. IX, 159 Seiten. 1973. DM 18,–

Vol. 336: L'Analyse Harmonique dans le Domaine Complexe. Edité par E. J. Akutowicz. VIII, 169 pages. 1973. DM 18,–

Vol. 337: Cambridge Summer School in Mathematical Logic. Edited by A. R. D. Mathias and H. Rogers. IX, 660 pages. 1973. DM 42,–

Vol: 338: J. Lindenstrauss and L. Tzafriri, Classical Banach Spaces. IX, 243 pages. 1973. DM 22,–

Vol. 339: G. Kempf, F. Knudsen, D. Mumford and B. Saint-Donat, Toroidal Embeddings I. VIII, 209 pages. 1973. DM 20,–

Vol. 340: Groupes de Monodromie en Géométrie Algébrique. (SGA 7 II). Par P. Deligne et N. Katz. X, 438 pages. 1973. DM 40,–

Vol. 341: Algebraic K-Theory I, Higher K-Theories. Edited by H. Bass. XV, 335 pages. 1973. DM 26,–

Vol. 342: Algebraic K-Theory II, "Classical" Algebraic K-Theory, and Connections with Arithmetic. Edited by H. Bass. XV, 527 pages. 1973. DM 36,–

Vol. 343: Algebraic K-Theory III, Hermitian K-Theory and Geometric Applications. Edited by H. Bass. XV, 572 pages. 1973. DM 38,–

Vol. 344: A. S. Troelstra (Editor), Metamathematical Investigation of Intuitionistic Arithmetic and Analysis. XVII, 485 pages. 1973. DM 34,–

Vol. 345: Proceedings of a Conference on Operator Theory. Edited by P. A. Fillmore. VI, 228 pages. 1973. DM 20,–

Vol. 346: Fučík et al., Spectral Analysis of Nonlinear Operators. II, 287 pages. 1973. DM 26,–

Vol. 347: J. M. Boardman and R. M. Vogt, Homotopy Invariant Algebraic Structures on Topological Spaces. X, 257 pages. 1973. DM 22,–

Vol. 348: A. M. Mathai and R. K. Saxena, Generalized Hypergeometric Functions with Applications in Statistics and Physical Sciences. VII, 314 pages. 1973. DM 26,–

Vol. 349: Modular Functions of One Variable II. Edited by W. Kuyk and P. Deligne. V, 598 pages. 1973. DM 38,–

Vol. 350: Modular Functions of One Variable III. Edited by W. Kuyk and J.-P. Serre. V, 350 pages. 1973. DM 26,–

Vol. 351: H. Tachikawa, Quasi-Frobenius Rings and Generalizations. XI, 172 pages. 1973. DM 18,–

Vol. 352: J. D. Fay, Theta Functions on Riemann Surfaces. V, 137 pages. 1973. DM 16,–

Vol. 353: Proceedings of the Conference on Orders, Group Rings and Related Topics. Organized by J. S. Hsia, M. L. Madan and T. G. Ralley. X, 224 pages. 1973. DM 20,–

Vol. 354: K. J. Devlin, Aspects of Constructibility. XII, 240 pages. 1973. DM 22,–

Vol. 355: M. Sion, A Theory of Semigroup Valued Measures. V, 140 pages. 1973. DM 16,–

Vol. 356: W. L. J. van der Kallen, Infinitesimally Central-Extensions of Chevalley Groups. VII, 147 pages. 1973. DM 16,–

Vol. 357: W. Borho, P. Gabriel und R. Rentschler, Primideale in Einhüllenden auflösbarer Lie-Algebren. V, 182 Seiten. 1973. DM 18,–

Vol. 358: F. L. Williams, Tensor Products of Principal Series Representations. VI, 132 pages. 1973. DM 16,–

Vol. 359: U. Stammbach, Homology in Group Theory. VIII, 183 pages. 1973. DM 18,–

Vol. 360: W. J. Padgett and R. L. Taylor, Laws of Large Numbers for Normed Linear Spaces and Certain Fréchet Spaces. VI, 111 pages. 1973. DM 16,–

Vol. 361: J. W. Schutz, Foundations of Special Relativity: Kinematic Axioms for Minkowski Space Time. XX, 314 pages. 1973. DM 26,–

Vol. 362: Proceedings of the Conference on Numerical Solution of Ordinary Differential Equations. Edited by D. Bettis. VIII, 490 pages. 1974. DM 34,–

Vol. 363: Conference on the Numerical Solution of Differential Equations. Edited by G. A. Watson. IX, 221 pages. 1974. DM 20,–

Vol. 364: Proceedings on Infinite Dimensional Holomorphy. Edited by T. L. Hayden and T. J. Suffridge. VII, 212 pages. 1974. DM 20,–

Vol. 365: R. P. Gilbert, Constructive Methods for Elliptic Equations. VII, 397 pages. 1974. DM 26,–

Vol. 366: R. Steinberg, Conjugacy Classes in Algebraic Groups (Notes by V. V. Deodhar). VI, 159 pages. 1974. DM 18,–

Vol. 367: K. Langmann und W. Lütkebohmert, Cousinverteilungen und Fortsetzungssätze. VI, 151 Seiten. 1974. DM 16,–

Vol. 368: R. J. Milgram, Unstable Homotopy from the Stable Point of View. V, 109 pages. 1974. DM 16,–

Vol. 369: Victoria Symposium on Nonstandard Analysis. Edited by A. Hurd and P. Loeb. XVIII, 339 pages. 1974. DM 26,–

Vol. 370: B. Mazur and W. Messing, Universal Extensions and One Dimensional Crystalline Cohomology. VII, 134 pages. 1974. DM 16,–

Vol. 371: V. Poenaru, Analyse Différentielle. V, 228 pages. 1974. DM 20,–

Vol. 372: Proceedings of the Second International Conference on the Theory of Groups 1973. Edited by M. F. Newman. VII, 740 pages. 1974. DM 48,–

Vol. 373: A. E. R. Woodcock and T. Poston, A Geometrical Study of the Elementary Catastrophes. V, 257 pages. 1974. DM 22,–

Vol. 374 S. Yamamuro, Differential Calculus in Topological Linear Spaces. IV, 179 pages. 1974. DM 18,–

Vol. 375: Topology Conference 1973. Edited by R. F. Dickman Jr. and P. Fletcher. X, 283 pages. 1974. DM 24,–

Vol. 376: D. B. Osteyee and I. J. Good, Information, Weight of Evidence, the Singularity between Probability Measures and Signal Detection. XI, 156 pages. 1974. DM 16.–

Vol. 377: A. M. Fink, Almost Periodic Differential Equations. VIII, 336 pages. 1974. DM 26,–

Vol. 378: TOPO 72 – General Topology and its Applications. Proceedings 1972. Edited by R. Alò, R. W. Heath and J. Nagata. XIV, 651 pages. 1974. DM 50,–

Vol. 379: A. Badrikian et S. Chevet, Mesures Cylindriques, Espaces de Wiener et Fonctions Aléatoires Gaussiennes. X, 383 pages. 1974. DM 32,–

Vol. 380: M. Petrich, Rings- and Semigroups. VIII, 182 pages 1974. DM 18,–

Vol. 381: Séminaire de Probabilités VIII. Edité par P. A. Meyer IX, 354 pages. 1974. DM 32,–

Vol. 382: J. H. van Lint, Combinatorial Theory Seminar Eindhoven University of Technology. VI, 131 pages. 1974. DM 18,–

Vol. 383: Séminaire Bourbaki – vol. 1972/73. Exposés 418-435 IV, 334 pages. 1974. DM 30,–

Vol. 384: Functional Analysis and Applications, Proceedings 1972. Edited by L. Nachbin. V, 270 pages. 1974. DM 22,–

Vol. 385: J. Douglas Jr. and T. Dupont, Collocation Methods fo Parabolic Equations in a Single Space Variable (Based on C^1 Piecewise-Polynomial Spaces). V, 147 pages. 1974. DM 16,–

Vol. 386: J. Tits, Buildings of Spherical Type and Finite BN Pairs. IX, 299 pages. 1974. DM 24,–

Vol. 387: C. P. Bruter, Eléments de la Théorie des Matroïdes V, 138 pages. 1974. DM 18,–

Vol. 388: R. L. Lipsman, Group Representations. X, 166 pages 1974. DM 20,–

Vol. 389: M.-A. Knus et M. Ojanguren, Théorie de la Descent et Algèbres d' Azumaya. IV, 163 pages. 1974. DM 20,–

Vol. 390: P. A. Meyer, P. Priouret et F. Spitzer, Ecole d'Eté d Probabilités de Saint-Flour III – 1973. Edité par A. Badrikia et P.-L. Hennequin. VIII, 189 pages. 1974. DM 20,–

Vol. 391: J. Gray, Formal Category Theory: Adjointness for 2 Categories. XII, 282 pages. 1974. DM 24,–

Vol. 392: Géométrie Différentielle, Colloque, Santiago d Compostela, Espagne 1972. Edité par E. Vidal. VI, 225 page 1974. DM 20,–

Vol. 393: G. Wassermann, Stability of Unfoldings. IX, 164 page 1974. DM 20,–

Vol. 394: W. M. Patterson 3rd, Iterative Methods for the Solutic of a Linear Operator Equation in Hilbert Space – A Surve III, 183 pages. 1974. DM 20,–

Vol. 395: Numerische Behandlung nichtlinearer Integrodifferer tial- und Differentialgleichungen. Tagung 1973. Herausgegebe von R. Ansorge und W. Törnig. VII, 313 Seiten. 1974. DM 28,

Vol. 396: K. H. Hofmann, M. Mislove and A. Stralka, The Pontr agin Duality of Compact O-Dimensional Semilattices and i Applications. XVI, 122 pages. 1974. DM 18,–

Vol. 397: T. Yamada, The Schur Subgroup of the Brauer Grou V, 159 pages. 1974. DM 18,–

Vol. 398: Théories de l'Information, Actes des Rencontres d Marseille-Luminy, 1973. Edité par J. Kampé de Fériet et C. Picar XII, 201 pages. 1974. DM 23,–

Vol. 399: Functional Analysis and its Applications, Proceedings 973. Edited by H. G. Garnir, K. R. Unni and J. H. Williamson. XVII, 569 pages. 1974. DM 44,-

ol. 400: A Crash Course on Kleinian Groups - San Francisco 974. Edited by L. Bers and I. Kra. VII, 130 pages. 1974. DM 18,-

ol. 401: F. Atiyah, Elliptic Operators and Compact Groups. , 93 pages. 1974. DM 18,-

ol. 402: M. Waldschmidt, Nombres Transcendants. VIII, 277 ages. 1974. DM 25,-

ol. 403: Combinatorial Mathematics - Proceedings 1972. Edited y D. A. Holton. VIII, 148 pages. 1974. DM 18,-

ol. 404: Théorie du Potentiel et Analyse Hormonique. Edité par Faraut. V, 245 pages. 1974. DM 25,-

ol. 405: K. Devlin and H. Johnsbråten, The Souslin Problem. III, 132 pages. 1974. DM 18,-

ol. 406: Graphs and Combinatorics - Proceedings 1973. Edited y R. A. Bari and F. Harary. VIII, 355 pages. 1974. DM 30,-

ol. 407: P. Berthelot, Cohomologie Cristalline des Schémas de aracteristique p > o. VIII, 598 pages. 1974. DM 44,-

ol. 408: J. Wermer, Potential Theory. VIII, 146 pages. 1974. M 18,-

ol. 409: Fonctions de Plusieurs Variables Complexes, Séminaire rançois Norguet 1970-1973. XIII, 612 pages. 1974. DM 47,-

ol. 410: Séminaire Pierre Lelong (Analyse) Année 1972-1973. I, 181 pages. 1974. DM 18,-

ol. 411: Hypergraph Seminar. Ohio State University, 1972. dited by C. Berge and D. Ray-Chaudhuri. IX, 287 pages. 1974. M 28,-

ol. 412: Classification of Algebraic Varieties and Compact omplex Manifolds. Proceedings 1974. Edited by H. Popp. V, 33 pages. 1974. DM 30,-

ol. 413: M. Bruneau, Variation Totale d'une Fonction. XIV, 332 ages. 1974. DM 30,-

ol. 414: T. Kambayashi, M. Miyanishi and M. Takeuchi, Uniotent Algebraic Groups. VI, 165 pages. 1974. DM 20,-

ol. 415: Ordinary and Partial Differential Equations, Proceedings f the Conference held at Dundee, 1974. XVII, 447 pages. 1974. M 37,-

ol. 416: M. E. Taylor, Pseudo Differential Operators. IV, 155 ages. 1974. DM 18,-

ol. 417: H. H. Keller, Differential Calculus in Locally Convex paces. XVI, 131 pages. 1974. DM 18,-

ol. 418: Localization in Group Theory and Homotopy Theory nd Related Topics Battelle Seattle 1974 Seminar. Edited by . J. Hilton. VI, 171 pages. 1974. DM 20,-

ol. 419: Topics in Analysis - Proceedings 1970. Edited by . E. Lehto, I. S. Louhivaara, and R. H. Nevanlinna. XIII, 391 pages. 974. DM 35,-

ol. 420: Category Seminar. Proceedings, Sydney Category heory Seminar 1972/73. Edited by G. M. Kelly. VI, 375 pages. 974. DM 32,-

ol. 421: V. Poénaru, Groupes Discrets. VI, 216 pages. 1974. M 23,-

ol. 422: J.-M. Lemaire, Algèbres Connexes et Homologie des spaces de Lacets. XIV, 133 pages. 1974. DM 23,-

ol. 423: S. S. Abhyankar and A. M. Sathaye, Geometric Theory Algebraic Space Curves. XIV, 302 pages. 1974. DM 28,-

ol. 424: L. Weiss and J. Wolfowitz, Maximum Probability stimators and Related Topics. V, 106 pages. 1974. DM 18,-

ol. 425: P. R. Chernoff and J. E. Marsden, Properties of Infinite imensional Hamiltonian Systems. IV, 160 pages. 1974. DM 20,-

ol. 426: M. L. Silverstein, Symmetric Markov Processes. IX, 37 pages. 1974. DM 28,-

ol. 427: H. Omori, Infinite Dimensional Lie Transformation roups. XII, 149 pages. 1974. DM 18,-

ol. 428: Algebraic and Geometrical Methods in Topology, Proeedings 1973. Edited by L. F. McAuley. XI, 280 pages. 1974. M 28,-

Vol. 429: L. Cohn, Analytic Theory of the Harish-Chandra C-Function. IV, 154 pages. 1974. DM 18,-

Vol. 430: Constructive and Computational Methods for Differential and Integral Equations. Symposium 1974. Edited by D. L. Colton and R. P. Gilbert. VII, 476 pages. 1974. DM 37,-

Vol. 432: R. P. Pflug, Holomorphiegebiete, pseudokonvexe Gebiete und das Levi-Problem. VI, 210 Seiten. 1975. DM 23,-

Vol. 433: W. G. Faris, Self-Adjoint Operators. VII, 115 pages. 1975. DM 18,-

Vol. 434: P. Brenner, V. Thomée, and L. B. Wahlbin, Besov Spaces and Applications to Difference Methods for Initial Value Problems. II, 154 pages. 1975. DM 18,-

Vol. 435: C. F. Dunkl and D. E. Ramirez, Representations of Commutative Semitopological Semigroups. VI, 181 pages. 1975. DM 20,-